SIGMA 1

INVESTIGATING MATHEMATICS

David Kent and Keith Hedger

with David Burghes and Don Steward

HODDER AND STOUGHTON
LONDON SYDNEY AUCKLAND TORONTO

Other books in the SIGMA series:
SIGMA 1 Teacher's Book
SIGMA 2 Investigating Mathematics
SIGMA 2 Teacher's Book

British Library Cataloguing in Publication Data

Kent, David
SIGMA: investigating mathematics.
1
1. Mathematics – For schools
I. Title II. Hedger, Keith
510

ISBN 0 340 39760 8

First published 1988

Phototypeset by Gecko Limited, Bicester, Oxon
Printed in Great Britain
for Hodder and Stoughton Educational
a division of Hodder and Stoughton Ltd, Mill Road
Dunton Green, Sevenoaks, Kent by
Butler & Tanner Ltd, Frome

Preface

This book is part of the SIGMA Project, **Studies and Investigations into Generating Mathematical Activities**, which aims to enhance mathematical teaching and learning in school.

The project involves a range of activities, the aim of which is to provide material and suggest a method of approach which enriches the mathematics curriculum, and provokes discussion and the use of a wide range of equipment and resources.

SIGMA 1 is primarily intended for the 11 to 14 year old age group but some of the material may be appropriate for older and younger pupils. Similarly, SIGMA 2 is primarily intended for the 14 to 16 year old age group but its use need not be restricted to this age range.

The Teacher's Books which covers SIGMA 1 and SIGMA 2 indicate clearly our suggested approach for using the material, and it is expected that staff and pupils will wish to extend topics by adding their own material and ideas.

The common core activities take a particular mathematical theme and give ideas for motivating pupils and developing the topic. The investigations introduce, develop and give possible extensions for particular mathematical problems, whilst the applications develop the use of mathematical analysis applied to practically based problems.

In all of the activities we encourage the teaching of mathematics through discussion, investigative techniques, applications and problem solving — alongside the traditional exposition and practice of routines. The activities have been written so that they can be readily incorporated into an existing mathematics curriculum, SIGMA 2 has topics which can be incorporated into a GCSE coursework assessment scheme for any examining board.

The project was originally financed by the Southern Regional Examination Board.

Project Directors:

David Burghes Professor of Education, University of Exeter

Keith Hedger Mathematics Adviser, Shropshire LEA

David Kent Mathematics Adviser, Tameside LEA
(formerly Mathematics Adviser, Suffolk LEA)

Don Steward Head of Mathematics, Oldbury Wells School,
Bridgnorth, Shropshire

Contents

A Guide for Students

There is a great deal more to mathematics than merely doing sums or simply following a set of instructions. You will find that we have provided you with a number of situations to **investigate** — that is, take a good, long and, we hope, intelligent look at. From your investigation we hope that you will be able to make some observations and record your results. *Do not be afraid* to record results which may ultimately be discarded for some reason or another. The fear of getting something wrong often puts people off mathematics, even though some of the greatest mathematicians have made observations and put forward results which other people have disputed and even disproved. Often getting something wrong will lead you to the right answer. But mathematics is not so simple that things are either always right or wrong. So if you have an observation or a point of view to offer, *make it known.*

Having made some observations and recorded some results you are often asked to make a **generalisation** and sometimes to **prove** or **explain** the result. One of the chapters is called Generalisations and includes work on proofs. It is a very important chapter. As an introduction to some of these words and to give you some idea of what to expect, here is a very simple example. (If instead of looking at this example, you would like to see something more challenging, then take a look at the Generalisations chapter.)

Suppose we were looking at the lamp posts down one side of the street. We might see 4 lamp posts as:

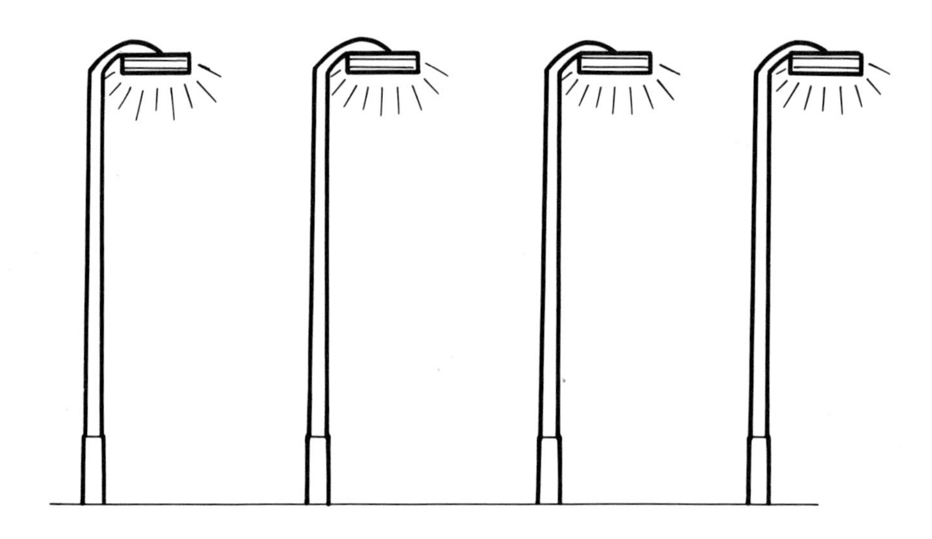

Now suppose we were asked how many spaces there were between the lamp posts. Well that is easy: there are 3 spaces.

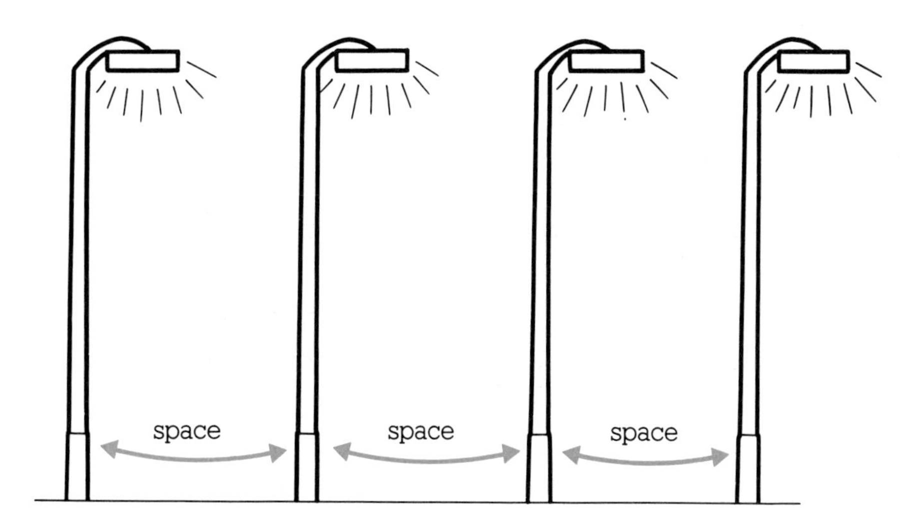

But now we carry the problem further and ask the question:
What is the general relationship between the number of lamp posts and the number of spaces between them?

Well of course the answer is very simple and you can probably see it already. But suppose that you did not, then one way of setting about it might be to look at a few simple cases, such as:

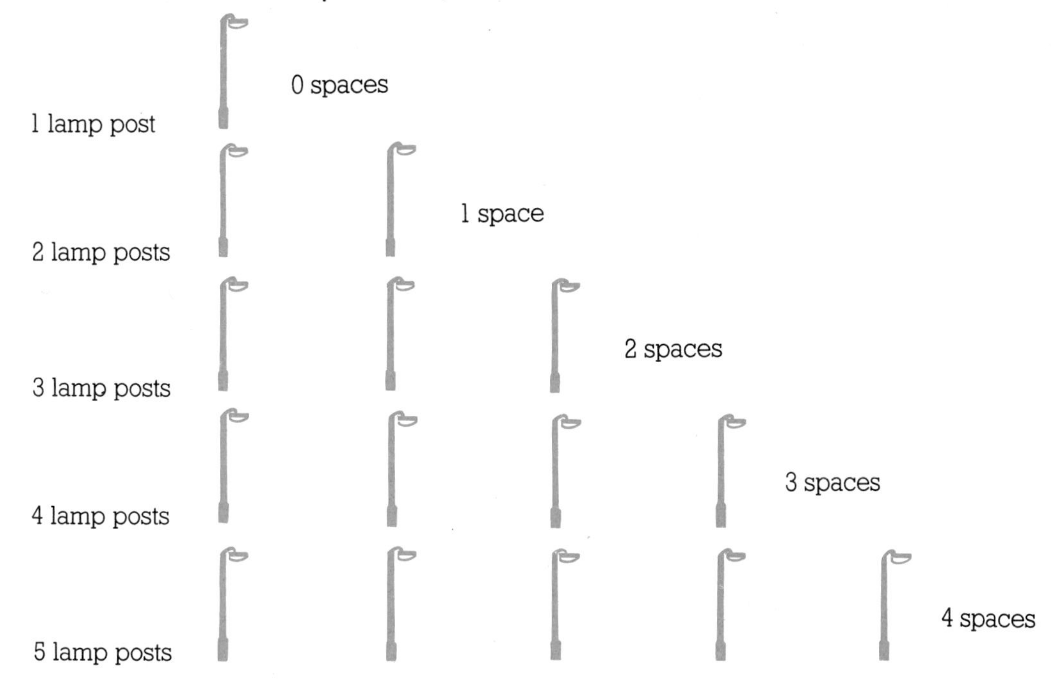

From these simple observations you could set up a table of results like this:

number of lamp posts	*number of spaces*
1	0
2	1
3	2
4	3
5	4

From these results together with a little bit of thought it seems reasonable to suggest that:

the number of spaces is always 1 *less* than the number of lamp posts

This is called a *generalisation*, or a *general statement*, or you say you have *generalised*.

What we have done, strictly, is to make a **conjecture** — which is really an educated guess based on our observations of the results.

We can obviously test our conjecture by looking at 6, 7, or more lamp posts, and if we do we will see that it works. Had it failed, then it would have been found to be deficient or incorrect in these circumstances. We would then say that the conjecture is **refuted**.

Mathematicians often express a generalised result in **symbolic form** using letters instead of expressing it in words as we have done. This can be tricky, especially if you have never seen it before. We will try to explain, but you may want to talk about this with your teacher.

Suppose we use a letter to represent the number of lamp posts. Let us say that we have:

n lamp posts

then, the number of spaces is 1 less than n. This is written as

$n - 1$ (not m or anything else)

We might express our generalisation now as:

number of lamp posts $= n$
number of spaces $= n - 1$

Having now made this generalisation and expressed it in symbolic form, we need to prove it. That is to say, we need to offer some explanation as to why it always works. These proofs, even in simple cases, are often quite difficult. They can be the trickiest part of a mathematical activity.

One way of looking at this problem, which leads to a proof, is to say that there is a sequence of:

Lamp post then space, lamp post then space, ... lamp post.

So if we write L for lamp post and S for space, it goes

LS, LS, LS, LS, LS, ..., LS, L

and in this sequence, the number of L's is 1 more than the number of S's. So if we have n of the letter L than we have $n - 1$ of the letter S. That completes the proof.

Perhaps the only two things we have not done in this simple example are to record any false observations and to offer any extensions to the problem. The false or deficient observations are difficult to make in this example because the problem is too simple but don't be worried if you do make false observations when you try problems for yourself. The extensions to the problem could be something about not having the lamp posts in a straight line, or looking at lamp posts on both sides of the street.

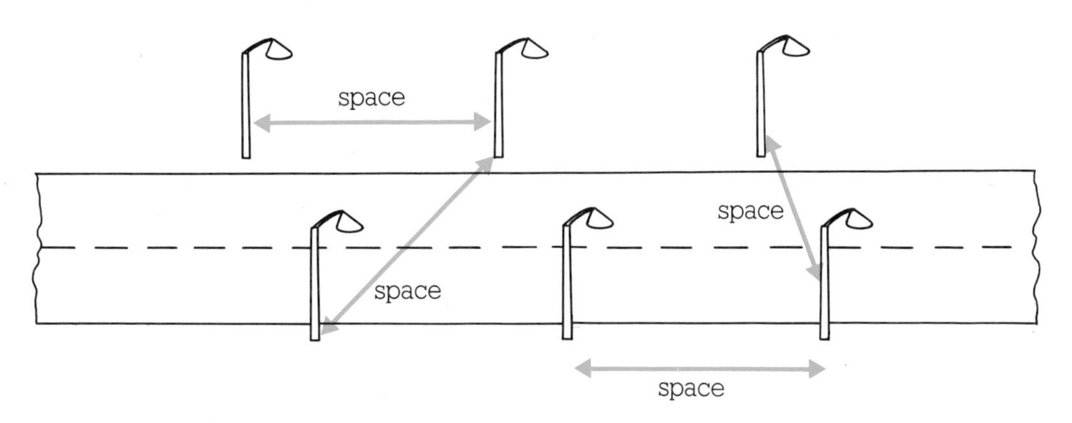

As we said earlier, this is a very simple example but we hope that it shows you something about how to approach other activities in this book which will be a challenge to you. We trust that you, your classmates and teachers will enjoy the challenge presented.

1 Closed Doodles

This is a **closed doodle**. All closed doodles *finish* at the point where they *start*. (It is like drawing a single loop.)

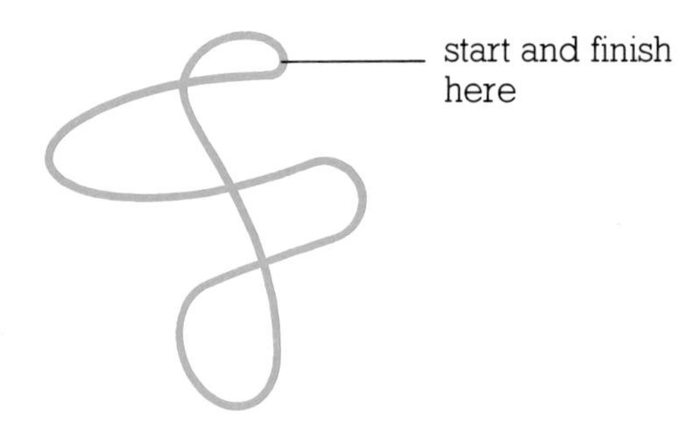

This doodle has 3 crossover points, 4 enclosed regions and 6 arcs.

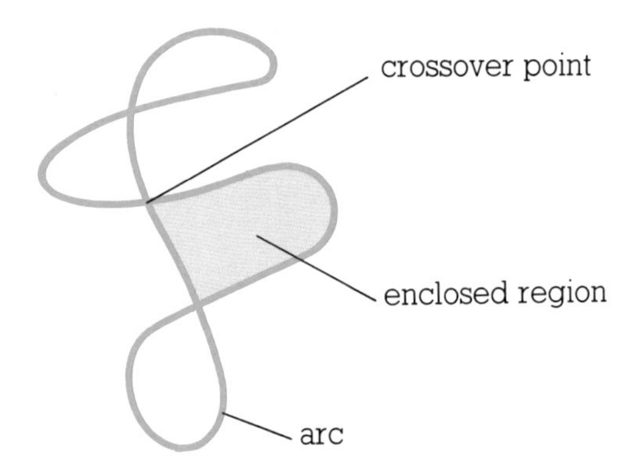

An arc is any branch of the doodle that starts at a crossover point and finishes at a crossover point.

Your task is to try to find the relationship between the number of crossovers, regions and arcs for any closed doodle.

Question 1

Draw some simple doodles like this:

What about this one?

Or what about clover leaf ones?

Draw at least 6 more closed doodles of your own. In each case, count the number of crossover points, enclosed regions and arcs. Make a table of your results.

Question 2

When you have done enough and think you have seen a pattern, write up your findings. In your writing give:

- □ a record of your results
- □ any diagrams
- □ any observations and patterns
- □ any explanations.

In addition, try to answer this question in your report:

Would it be possible to draw a closed doodle which has 18 crossover points; 26 regions; 50 arcs?

If your answer is *yes*, draw it. If your answer is *no*, say why.

Finally, try to make a **general rule** that is true for any closed doodle you draw, about the number of crossover points, enclosed regions and arcs in the doodle.

2 Euler's Formula
An extension to Closed Doodles

A cube has
8 vertices (corners)
6 faces
12 edges

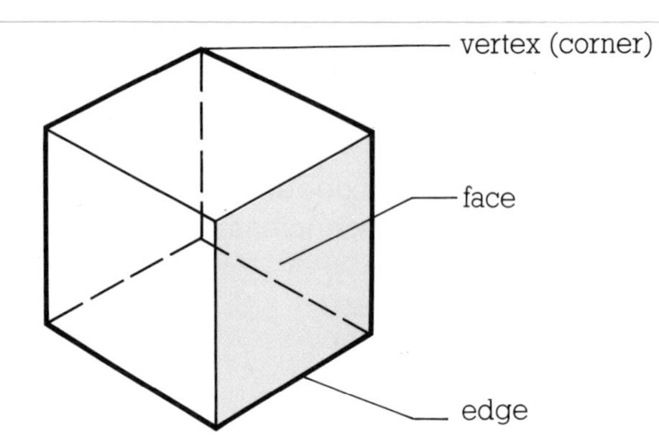

There is a relationship between the number of vertices, faces and edges for most well known solid shapes. This formula is known as **Euler's Formula** after the man who found it. He was a Swiss mathematician named Leonhard Euler, who lived from 1707 to 1783.

Your task is to work out this relationship between the number of *vertices, faces* and *edges.*

To help, you may wish to:
make some 3 dimensional models;
draw sketches of models;
use models already made or bought.

Question 1

Start by looking at some of these reasonably simple solids.

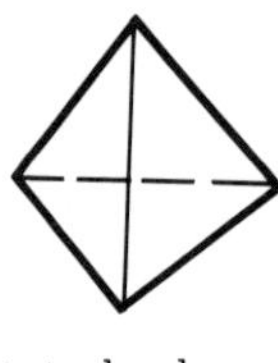
tetrahedron

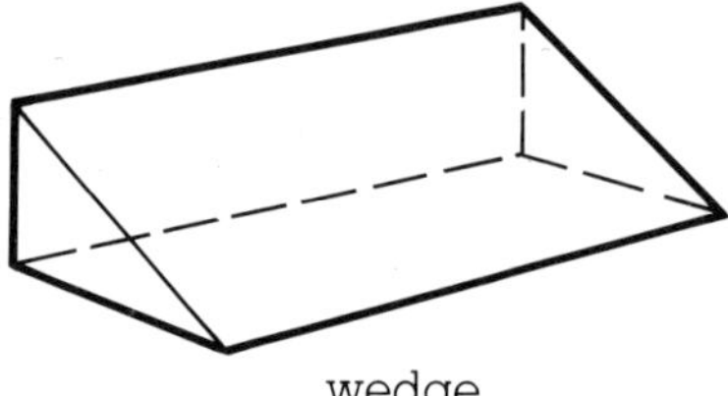
wedge

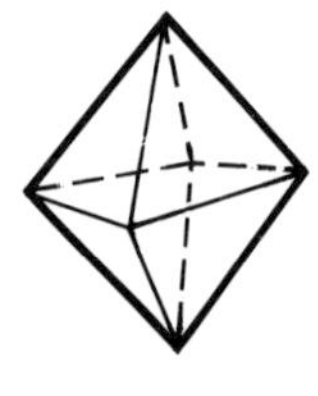
octahedron

Now look at some other shapes as well.

Question 2

Write a report on this work. In your report make sure you:

- □ record your results
- □ write down any observations
- □ explain any results
- □ **generalise** your results.

Question 3

If you wish to make a wall display about this work then please do so. Ask your teacher for suitable paper or other material.

3 Frogs

In this exercise you move the counters according to the rules shown below and try to finish with all the black counters on the right and all the white counters on the left.

starting position

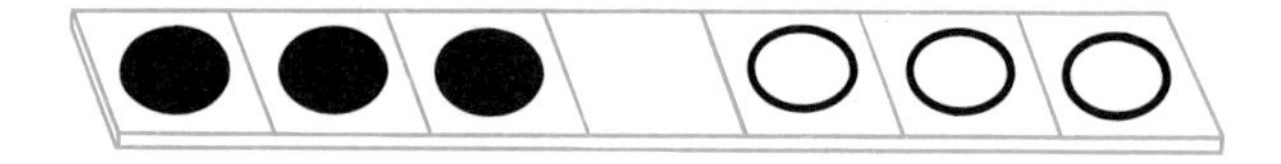

Rules

(i) move only one counter at a time,
(ii) black move only to the right,
white move only to the left,
(iii) a counter moves into an empty adjacent space,

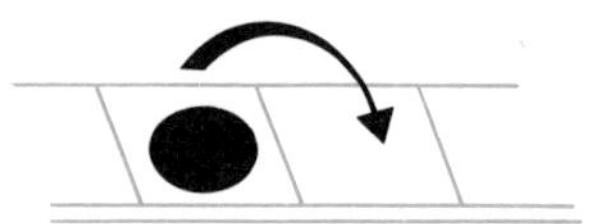

(iv) a counter can jump over one counter — but not more — of the other colour to land in an empty space.

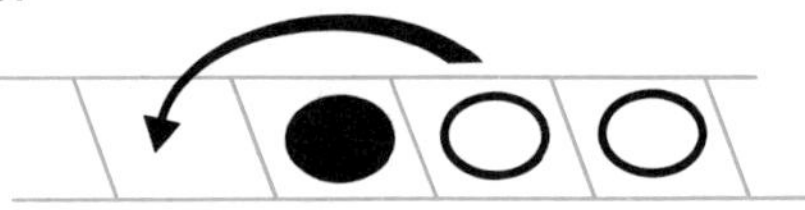

Can you finish like this?

As well as just being able to play the game, your task is to find a general relationship between the numbers of counters of each colour and the total number of moves needed to complete the game.

Question 1

After trying the game with 3 counters of each colour, try it with

1, 2, 4, 5, 6, . . .   counters of each colour.

Question 2

- □ Record your results in a table or similar.
- □ Write down any strategies you might have used.
- □ Write down any observations you can make.
- □ How many moves would it take if you had 15 counters on each side? State how you work out your result.
- □ Can you **generalise** your results?

Question 3

Now try the game where the numbers of counters of each colour are different.

Try some more of your own.

With the different numbers for the two colours, do as many examples as you need before you feel able to make a **generalisation**. Test your results if you wish.

How many moves would it take if you had 15 counters of one colour and 20 counters of another colour?

4 Charity Cycle Ride

A group of boys and girls from the youth club are taking part in a charity cycle ride. The rules are that they must start at the Sports' Centre, visit each of 4 churches and finish at the Sports' Centre. They have a map, showing the distances — in kilometres, to one decimal place — between the churches and from each church to the Sports' Centre. The map is not to scale.

The map is:

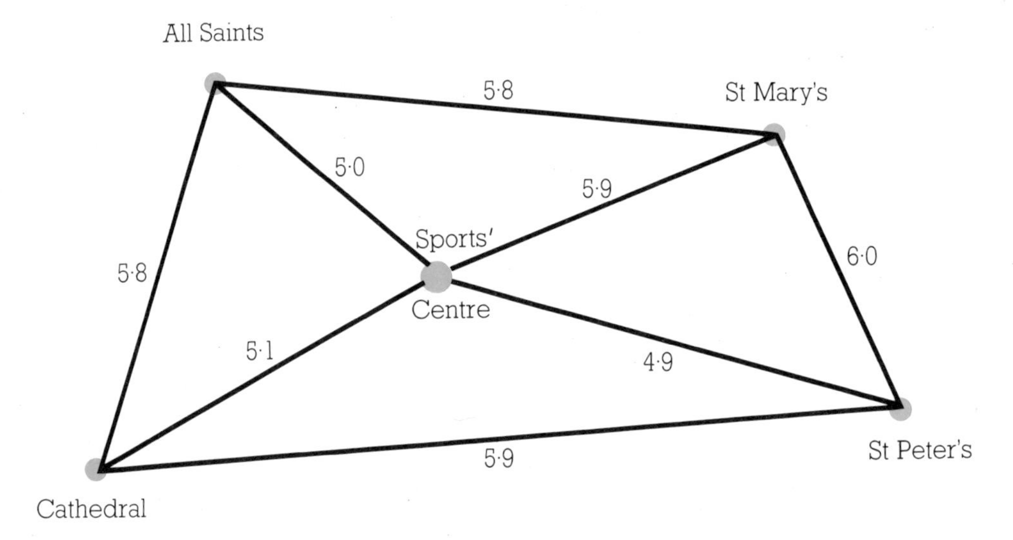

Question 1

Can you work out the route they should take so that the distance they have to cycle is as short as possible?

What is that minimum distance?

Question 2

Suppose that they cycle at an average speed of 20 kilometres per hour (km/h). How long, to the nearest minute, will it take them to ride the minimum distance?

Question 3

How long would it take if they did the maximum distance?

Question 4

Make up similar problems of your own, with your own distances, using

1 3 churches
2 5 churches
3 6 churches.

Ask your neighbours to try your problems and you try theirs. Make sure that you know or can work out the minimum distance in each case.

Question 5

You may need an Ordnance Survey or similar map for this project.

Create a problem like the one in this chapter but use real places. Choose something such as your own town, or your school and some real places to visit — nearby towns, local shops, other schools etc. Use actual distances, obtained from a map perhaps.

Find the minimum route yourself or set this as a problem for your class-mates.

Perhaps you could make a wall chart about it when you have finished the work.

5 Handshakes

A group of people meet for the first time at a party. They are told that everyone has to shake hands with everyone else just once.

Your task is to find a **generalised result** connecting the number of people at the party and the total number of handshakes.

Question 1

Try a few **simple cases** first. For instance, if there was only 1 person at the party there would be 0 handshakes. If there were 2 people at the party there would be 1 handshake.

Suppose that there were 3 people at the party, let us call them Mark, Clare and Steven. We would have 3 handshakes in all. One of these would be: Mark shakes with Clare. What are the other 2?

Question 2

Now do the same for 4 people, then 5 people, then 6 people and more if you wish. Give the people names if you like.

□ Record your results.

□ Form a table of values.

□ Write down any observations you make.

□ Can you **generalise** the result?

Question 3

What if there were lots of people at the party? Could you work out the total number of handshakes if there were:

1 30 people at the party
2 100 people at the party?

Question 4

Suppose that you were told how many handshakes there were, but not how many people there were. Could you work out the number of people at the party if there were:

1 66 handshakes
2 1225 handshakes?

6 Geoboard 1
Journeys

Using a geoboard or dotted paper, you can make 'journeys' like the ones shown here. At the end of each stage, you take a right turn through 90°. For a [2 6 3] journey, you move 2 spaces for the first stage, turn, move 6 spaces for the second stage, turn, move 3 spaces for the next stage and carry on until you reach the beginning again. Some journeys don't take you back to the beginning. Look at all the different examples shown and then do question 1.

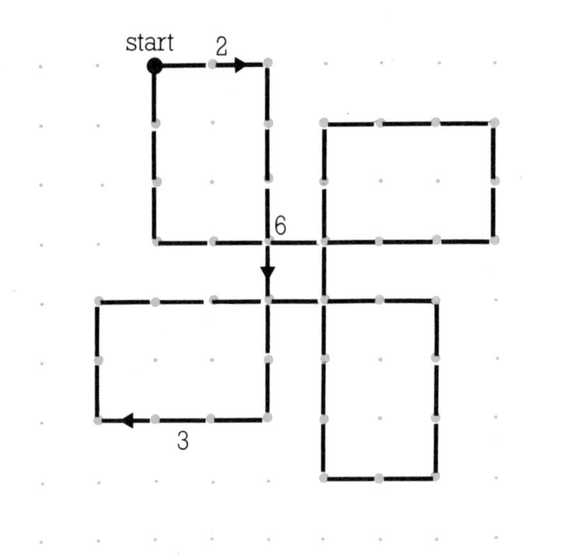

This is a [2 6 3] journey.

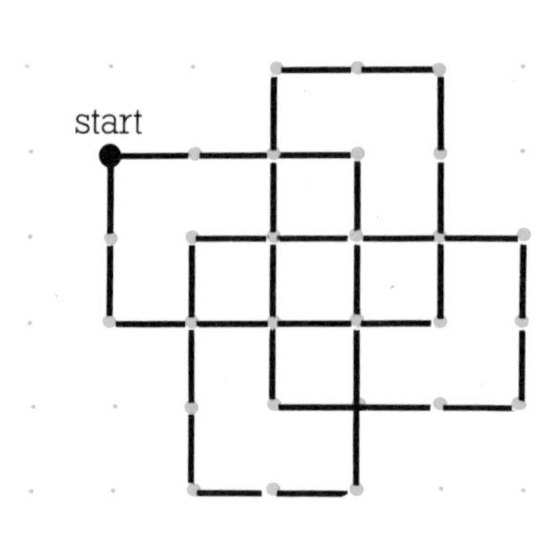

This is a [3 4 2].

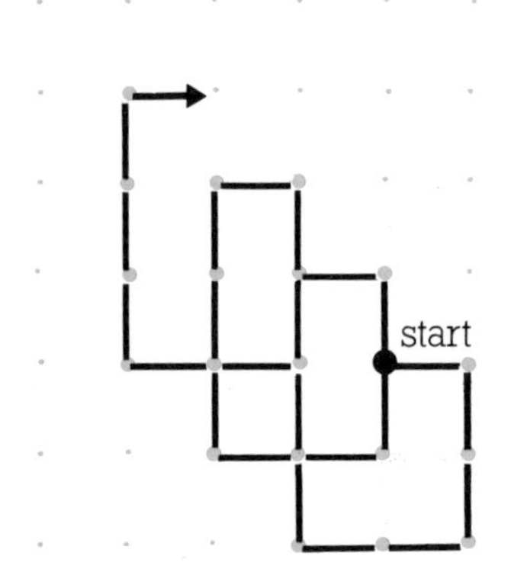

This one uses 4 numbers — a [1 2 2 3].

What happens if you use 0 or negative numbers?
For example, [1 2 0 3]

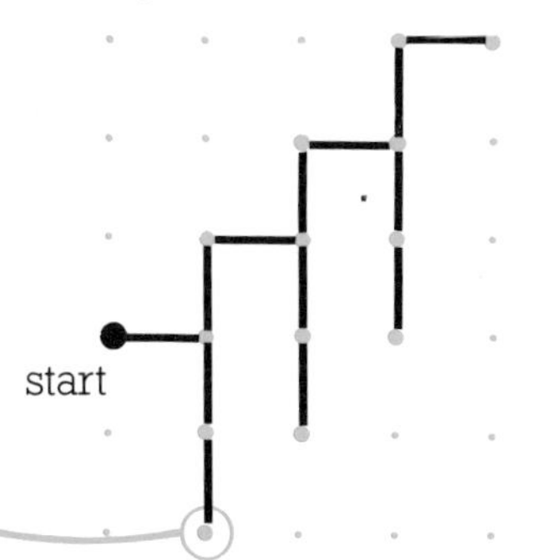

Having turned right following the '2' step, you go forward '0' before turning right again and travelling forward '3' steps.

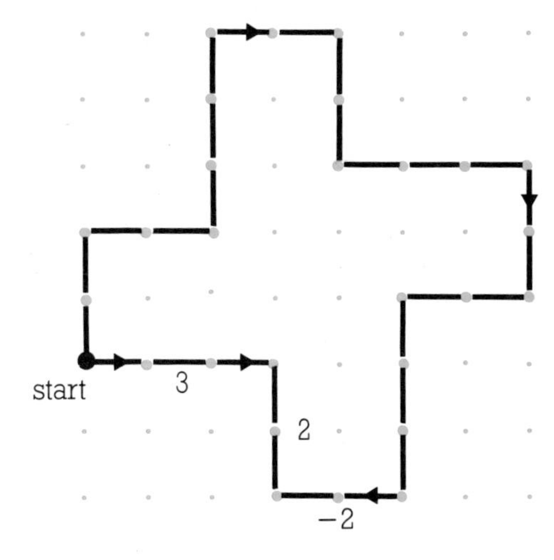

This is a [3 2 (−2)].

Question 1

Try some of your own 'journeys', experimenting with different numbers of stages, zeros and negative numbers. Collect your results together and make any possible observations.

Question 2

If you have access to LOGO on a computer you could try a version of the following program, (written here in Logotron LOGO).

```
TO WALK :S1 :S2 :S3
FD :S1 RT 90 FD :S2 RT 90 FD :S3 RT 90
WALK :S1 :S2 :S3
END
```

Using this program, WALK 20 60 30 (scaled to make it visible on the screen) should give you the result for [263] in the first example of this chapter.

Question 3

What happens when the angle of turn at the end of each stage is changed?

Try using triangular isometric dot paper or adjusting the program above to include the variable (:ANG) and changing (RT 90) to (RT :ANG).

7 Loci

The diagram shows a black counter, a grey counter and a set of white counters. The white counters are placed so that their distances from the black are equal to their distances from the grey. In each case WB = WG.

W

B G

The set of white counters all lie on a straight line — it is in fact the **perpendicular bisector** of the line from B to G. This line on which the white counters lie is called a **locus**. To be accurate, it is the locus of the points W, such that WB = WG.

Question 1

Work in pairs or groups of 3, with counters on the floor or another convenient place.

Use your counters to set up the locus for which

$$WB = 2WG$$

This means that the distance from the white to the black is 2 times the distance from the white to the grey. Black and grey should be about 15 – 20 cm apart. Make a sketch of this locus in your book or on paper. Write down any methods you use.

In the original example, the locus was a straight line. What is it in this case?

Question 2

Either using counters or by using points marked on paper, set up each of the following loci:

1 WB = 3WG

2 2WB = WG

3 WB = ½WG

4 WB = 5WG

You will have to be careful in your choice of distances between B and G. What observations can you make about these loci?

Question 3

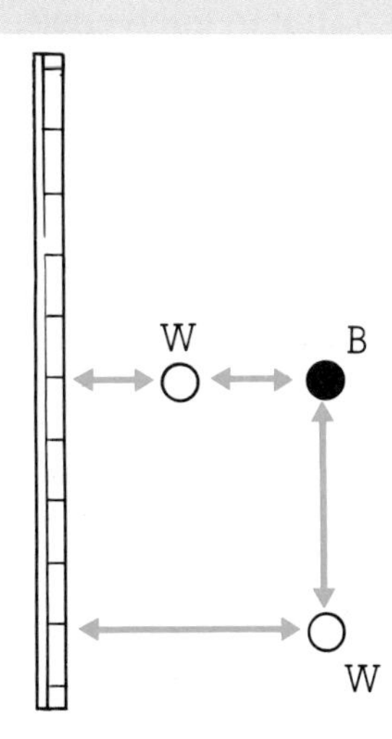

The diagram shows a picture of a metre ruler, black counter and 2 white counters.

Each of the white counters is placed so that its distance from the black counter is equal to its perpendicular distance from the metre ruler.

Again work in pairs or groups of 3 in a convenient place. Set up the situation with the counters and the metre ruler.

Place at least 6 more white counters so that their perpendicular distance from the metre ruler equals their distance from the black. Sketch the final situation in your book or on paper.

The locus in question 3 is called a **parabola**. It occurs when the distance between a moving point and a fixed point is equal to the perpendicular distance from the moving point to a straight line.

Question 4

In the diagram the black and grey counters are 10 cm apart. The white counter is placed so that:

WB + WG = 16cm

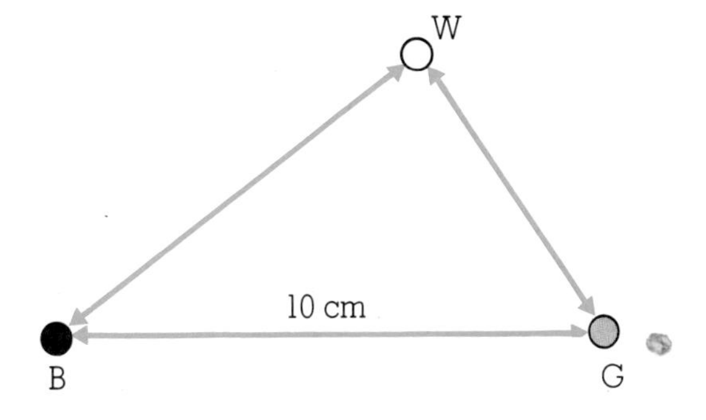

Set this situation up in a convenient place. Then put down more white counters so that in each case:

$$WB + WG = 16 \text{ cm}$$

Sketch the final situation on paper or in your book.

The locus in question 4 is called an **ellipse**. It occurs when the sum of the distances between two points and a moving point is constant.

Just like the parabola it occurs in many different places.
The loci which we now call

circle
parabola
ellipse

are all examples of **conic sections**.

In the next question we will show you why they have this name.

Conic sections

Question 5

You will require a transparent plastic model of a cone. There should be a hole in the base of the cone. You will also require a small amount of sand, just enough to fill about half the cone. When you have half-filled the cone, close the hole in the base. The idea now is that you shake the sand so that the sand level lies either

parallel with the base of the cone,

at an angle to the base,

parallel to a side of the cone.

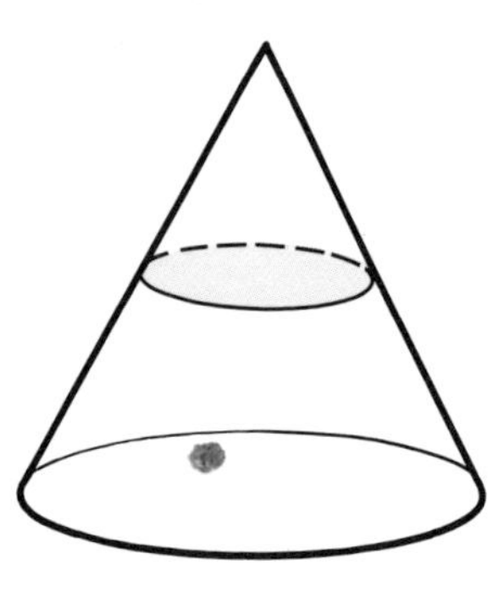

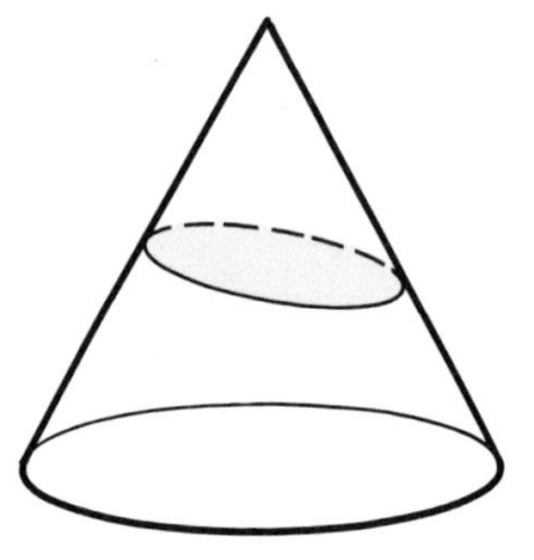

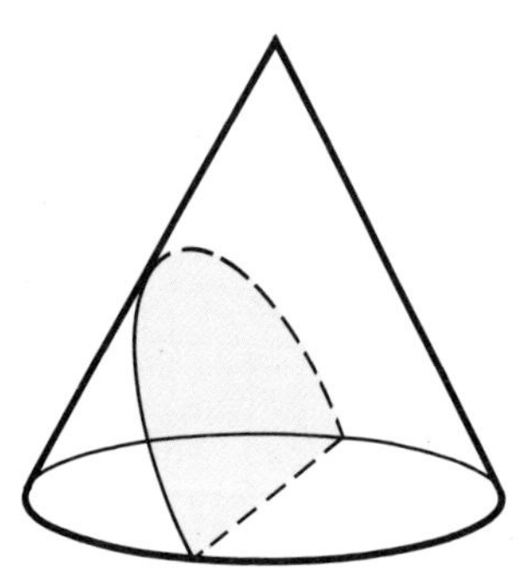

If you look at the sand, in each case, you should be able to see either a **circle** or an **ellipse** or a **parabola**.

Make sketches of these three situations with the cone.

The circle, ellipse and parabola are called conic sections because they are the shapes of the **cross section** when a cone is sliced. Can you see and explain this?

Question 6

Try to find some books or other written material on conic sections, curves or loci. Take time to read and make notes as you wish.

Question 7

In this picture we have two lines drawn at an angle. They are both marked from 1 to 12. We have joined 1 and 12, 2 and 11. Copy the picture and join all the pairs that add up to 13.

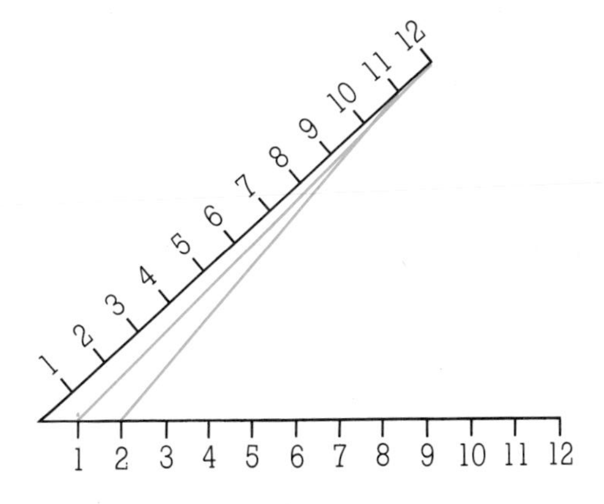

Your final picture should have something to do with the parabola. Can you see this? Can you explain it?

It may help to draw a few diagrams in answering these questions:

(i) Does the angle between the lines matter?
(ii) Does it matter that we use 12 graduation marks?
(iii) Does it matter whether or not the markings on the two lines are equally spaced?

Write a few notes on the answers to these questions.

8 Generalisations

Triangle patterns

A pattern of triangles is made from matchsticks

3 matches 1 triangle

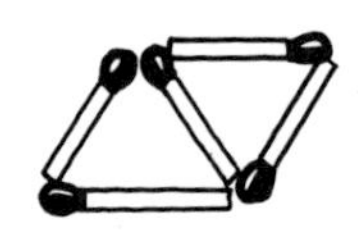

5 matches 2 triangles

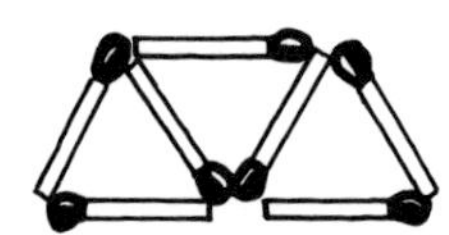

7 matches 3 triangles

A pattern suggests itself

number of triangles	*number of matches required*
1	3
2	5
3	7
4	9

We get the number on the right by:

multiplying the number on the left by 2 then adding on 1

We **generalise** this result by saying that:

the number of matches equals 2 times the number of triangles plus 1

Or in **symbolic form** we write
number of triangles $= n$
number of matches $= (2 \times n) + 1$
which is written
number of matches $= 2n + 1$

Our **generalisation** is
if number of triangles $= n$
number of matches $= 2n + 1$

This **hypothesis** is based on our observations of a number pattern. We could also call it a **conjecture**.

We now need to either **prove** the hypothesis or show it to be **false**. If it is false then we **refute** it. This generalisation is actually true, so we will prove it. Often we will not know that our generalisation is true. We sometimes have to make a good guess or have a feeling about it.

Our proof:

To make 1 triangle we need 3 matches

3 matches

To make 2 triangles we need another 2 matches

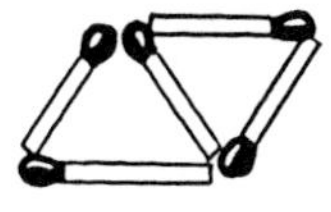

$3 + 2$ matches

To make 3 triangles we need another 2 matches

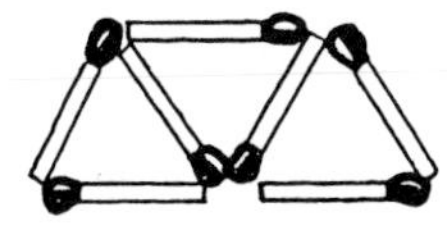

$3 + 2 + 2$
or $3 + (2 \times 2)$ matches

To make 4 triangles we need another 2 matches i.e.

4 triangles need	$3 + 2 + 2 + 2$
4 triangles need	$3 + (3 \times 2)$ matches

So the pattern is:

1 triangle	3 matches
2 triangles	$3 + 2$ matches
3 triangles	$3 + (2 \times 2)$ matches
4 triangles	$3 + (3 \times 2)$ matches
5 triangles	$3 + (4 \times 2)$ matches

and so

n triangles $3 + (n - 1) \times 2$ matches

but $(n - 1) \times 2 = 2n - 2$. Do you see why?

So n triangles need

$$3 + 2n - 2$$
$$= 2n + 1 \text{ matches}$$

Question 1

The picture shows 3 lamp posts.

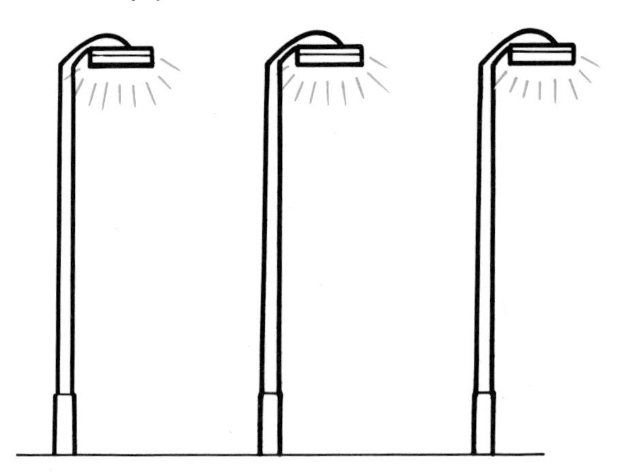

There are 2 spaces between them. Write down a **generalisation** for the number of spaces between *n* lamp posts.

Question 2

The first odd number is	1.
The second odd number is	3.
The third odd number is	5.

Write down a general result for the *n*th odd number.

Question 3

A fence is made with some posts connected by lengths of wire. In the picture we can see 4 posts and 9 lengths of wire.

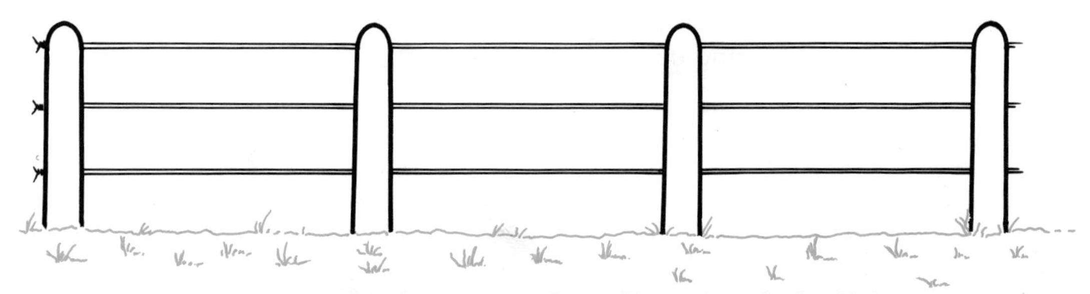

Write down a **general result** for the number of lengths of wire if there are *n* posts.

Prove your result.

Triangular numbers

Question 4

The **triangular numbers** follow a pattern.

first triangular	×	1
second triangular	×× ×	1 + 2
third triangular	××× ×× ×	1 + 2 + 3

So the pattern goes:

$$1 = 1 = \frac{1}{2}(2 \times 1)$$
$$1 + 2 = 3 = \frac{1}{2}(3 \times 2)$$
$$1 + 2 + 3 = 6 = \frac{1}{2}(4 \times 3)$$
$$1 + 2 + 3 + 4 = 10 = \frac{1}{2}(5 \times 4)$$

Generalise this pattern for the *n*th triangular number.

Square patterns

Question 5

A pattern of connected squares is made using matches. It looks like this:

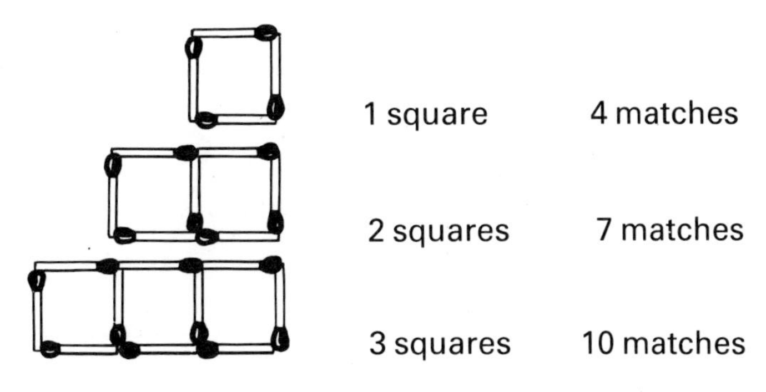

How many matches would you need to make

1 4 squares
2 5 squares
3 10 squares?

Generalise the result for the number of matches needed to make *n* squares. Prove this result.

Question 6

There is a famous connection between **odd** numbers and **square** numbers. It goes something like this:

$$1 = 1 = 1^2$$
$$1 + 3 = 4 = 2^2$$
$$1 + 3 + 5 = 9 = 3^2$$

Generalise this result.

Diagonals

Question 7

A diagonal is a line joining two corners of a 4 or more sided figure. (The diagonal cannot be one of the sides of the figure.)

A square has 4 corners and 2 diagonals.

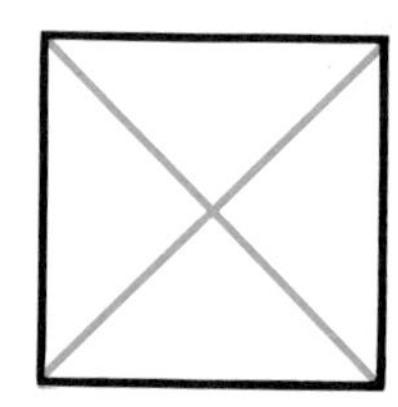

A pentagon has 5 corners and 5 diagonals.

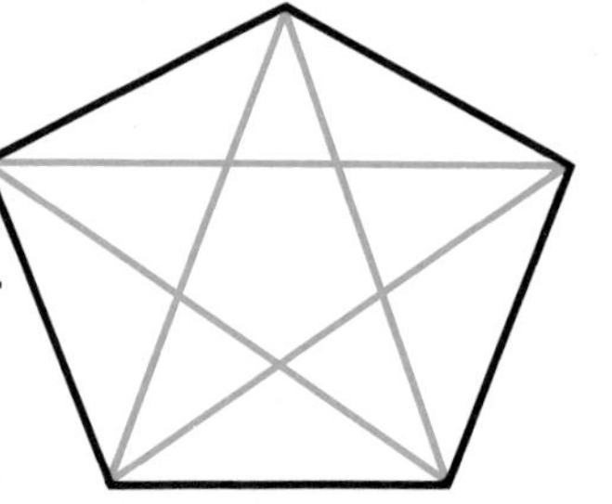

Generalise the result.

9 Cuisenaire Activity 1

w : *white*
r : *red*
g : *light green*
p : *pink*
y : *yellow*
d : *dark green*
b : *black*
t : *tan* (brown)
B : *blue*
O : *orange*

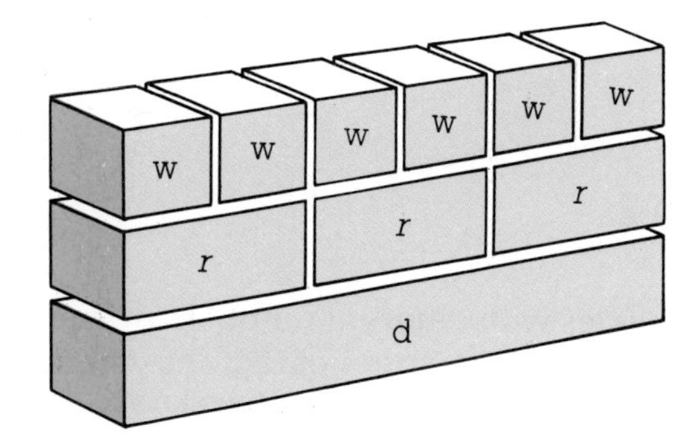

The list shows the 10 Cuisenaire rods. Each one is a different colour and length. The first rod (the *white* one) is 1 unit long, the second is 2 units long and so on up to the *orange* rod which is 10 units long.

The *yellow* rod is the same length as a **rod train** made from a *pink* rod and a *white* rod.

| *y* | = | *p* | *w* |

Question 1

(i) Find all 16 different rod trains equal in length to the *yellow* rod assuming that

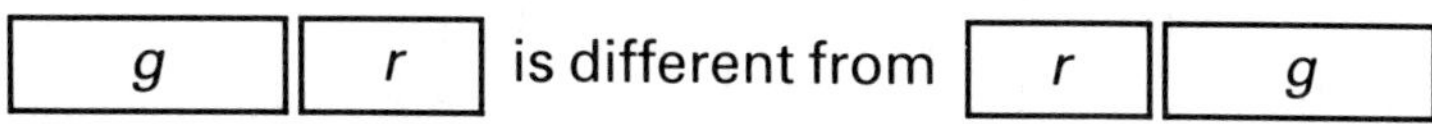

(ii) Now devise, and explain, a method for finding the number of rod trains equal in length to the:

1 *pink* rod
2 *dark green* rod
3 *red* rod
4 Each of the other rods

(iii) Look at your results and try to find the **general pattern**.

In the next question you have to join points around the outside of a circle with *straight lines* and create the *maximum number of* regions inside the circle. In this problem, you will have to be careful about jumping to conclusions.

> *Example*
>
> Using 4 points around the circle gives 8 regions.
>
> 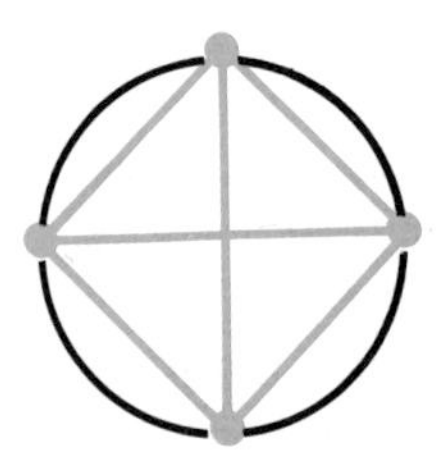

Question 2

Do this for circles with different numbers of points around the perimeter and in each case count the number of regions you make inside. Copy and fill in the following table. Try to find a **general pattern** in the results.

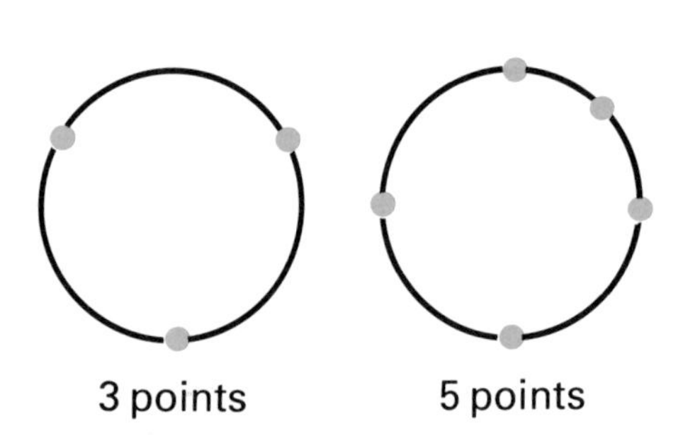

number of points	*number of regions*
0	
1	
2	
3	
4 →	8
5	
⋮	
n	

Rod algebra

In this question

[p][r] represents $(p + r)$.

Whereas $(p - r)$ is represented by this space:

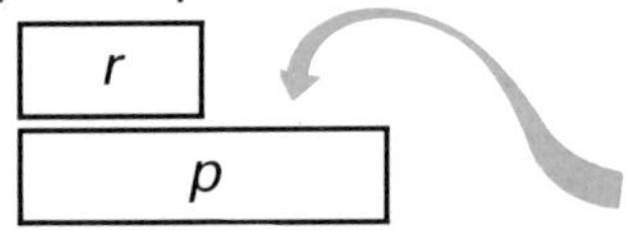

Question 3

Answer the following in terms of a single rod where possible:

1 $t - p = \square$
2 $g + r + y = \square$
3 $\square + r = y$
4 $y - \square = g$
5 $t = r + \square$
6 $O - t = \square$
7 $y + g - d = \square$
8 $\square = r + p$
9 $t - y = \square$
10 $t - \square + w = b$
11 $g + r = \square$
12 $y - r = \square$
13 $y - (r + r) = \square$
14 $(y + r) - g = \square$
15 $y - (r + \square) = r$
16 $y - g - \square = w$
17 $w + r + \square = p$
18 $t - (\square + w) = g$
19 $\square - (b + r) = w$
20 $p - g = \square$
21 $w + g + g + \square = B$
22 $2w + g = \square$
23 $w + 3g = \square$
24 $4w + 2g = \square$
25 $y + r = 2r + w + \square$
26 $d = \square + w$
27 $\square = g + d$
28 $B - (2r + p) = \square$
29 $y = 3w + \square$
30 $w + \square + y = b$
31 $O - 2r = \square$
32 $d + (b - 2g) = \square$
33 $w + r + \square + w = y$
34 $g + p = \square$
35 $B - b + r = \square$
36 $\square = O - (2r + g)$
37 $b - \square = r$
38 $d - \square = p$
39 $b - (w + \square + g) = r$
40 $3y - 2p = \square$
41 $B - 2\square = g$
42 $y - 4\square = w$
43 $O - 3\square = p$
44 $3y - 2(r + w) = \square$
45 $3y - 2r - 2w = \square$
46 $3y - 2r = \square + 2w$
47 $3y = \square + 2w + 2r$
48 $\frac{1}{2}$ of $(3y - \square) = r + w$

Question 4

Test the following statements (or equations) and identify whether they are true or false.

1 $r + g = g + r$
2 $w + r + g = r + w + g$
3 $3r = r + 2r$
4 $y - r = r - y$
5 $r + (p + y) = (r + p) + y$
6 $b - (r + w) = b - r + w$
7 $b - 2r = b - r - r$
8 $(b + y) - p = b + (y - p)$
9 $(t - p) - w = t - (p - w)$
10 $3y - 2p = (2y - p) + (y - p)$

Now make up true statements of your own which require brackets and a mixture of signs.

10 An Algebra of Actions

Question 1

You are given some actions, called action 1 and action 2. Decide, in each case, whether doing action 1 first then following it with action 2 gives the *same* result as action 2 followed by action 1.

	action 1	*action 2*
1	Put on your left shoe	Put on your left sock
2	Put on your shoes	Put on your watch
3	Comb your hair	Clean your teeth
4	Open the french window	Walk through the french window

Question 2

Make up 4 pairs of actions of your own, like those in question 1, such that:

(i) for the first two

action 1 followed by action 2 = action 2 followed by action 1

(ii) for the second two

action 1 followed by action 2 ≠ action 2 followed by action 1

Commutative relationships

The sentence

Peter is married to Jane

can be symbolised as

$$P \bullet J$$

where P = Peter; J = Jane; • = is married to.

It is also true that

Jane is married to Peter

which can be symbolised as

J•P

So in this case we have

P•J = J•P

which is called a **commutative** relationship.

There are other sentences which do not follow the commutative rule. For instance,

Sarah carries the bag

can be symbolised as

S•B

where S = Sarah; B = the bag; • = carries.
However in this case

S•B ≠ B•S

Question 3

Symbolise each of the following sentences, stating what the symbol represents. In each case say whether or not the general commutative rule holds.

1. Tom kicks the ball
2. Angela plays the piano
3. Louise is next to Karen
4. Stephen reads the book
5. Jill likes disco dancing
6. Fred is taller than Emma
7. Alan is the cousin of Pamela

Question 4

Make up 6 sentences of your own, all of which can be symbolised as X•Y. The first three should be commutative, so

X•Y = Y•X

The next three must *not* be commutative, so

X•Y ≠ Y•X

Distributive relationships

The sentence

George is thirty and married

can be symbolised as

$$G\bullet(T+M)$$

where G = George; T = thirty; M = married; • = is; + = and.
But surely this sentence is the *same* as saying

George is thirty and George is married

which can be symbolised as

$$(G\bullet T) + (G\bullet M)$$

Therefore in this case

$$G\bullet(T+M) = (G\bullet T) + (G\bullet M)$$

and we describe this by saying
'is' is **distributive** over 'and' or • is distributive over +.
If for any reason you think that the sentences

George is thirty and married

and

George is thirty and George is married

are not the same, then write down your reasons why.

Question 5

Johnson plays for England and Manchester United.

(i) Symbolise this sentence, using J = Johnson; E = England; M = Manchester United; • = to play for; + = and.
(ii) Does this sentence follow the distributive rule?

Question 6

Make up at least 4 sentences of your own, each of which follows the distributive rule.

Non-distributive relationships

Even with simple sentences which can be symbolised as

$$A \bullet (B + C)$$

some do not obey the distributive rule.

For instance, if T = this scarf; B = black; W = white; • = is; + = and, then

This scarf is black and white

can be symbolised as

$$T \bullet (B + W)$$

But,

$$(T \bullet B) + (T \bullet W)$$

is

This scarf is black and this scarf is white

The sentences 'This scarf is black and white' and 'This scarf is black and this scarf is white' do not have the same meaning.

The shorter sentence gives a mental picture of a single scarf which is black and white in colour (possibly striped). The longer sentence gives a mental picture of two scarves, one black and the other one white.

Perhaps your mental pictures are not the same. If this is the case, discuss it with your friends and see what they think.

Question 7

Make up a sentence which can be symbolised as

$$A \bullet (B + C)$$

but which does *not* obey the distributive rule, i.e.

$$A \bullet (B + C) \neq (A \bullet B) + (A \bullet C)$$

Give your reasons why the distributive rule does not work in this case.

Question 8

Make up two more sentences of the same type, where

$$A \bullet (B + C) \neq (A \bullet B) + (B \bullet C)$$

Again, give your reasons why the distributive rule does not work.

11 Line Segments

For 2 points, A and B, we have only 1 line segment AB.

For 3 points, A, B and C in a line we can have the line segments AB, BC and AC. So there are 3 line segments.

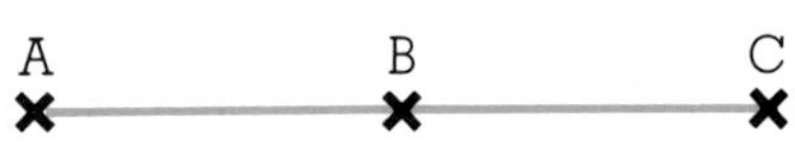

Question 1

Write down the full list of all line segments for the 4 points A, B, C and D in a line.

Be careful to *order* your process.

Question 2

Write down the full list of all line segments for the 5 points A, B, C, D, E all in a line.

Again be careful about ordering your process.

Question 3

Copy and complete the table. Extend your results up to the case where there are 10 points. Remember to put in the first result.

number of points	*number of line segments*
1	
2	1
3	3
4	
5	
⋮	

Question 4

What observations can you make about the number patterns which appear in the table? Discuss these with your neighbours.

Question 5

If there were

1 25
2 40
3 100

points in a line, how many line segments would there be in each case?
Explain how you get your results.

Question 6

Given the numbers of line segments

1 55
2 190
3 435

calculate how many points there would be.
Explain how you obtain the results.

Question 7

Generalise the result for the number of line segments when there are *n* points in a line.
Prove this or explain why the result is as it is.

12 Cuisenaire Activity 2
Fraction statements

w : *white*
r : *red*
g : *light green*
p : *pink*
y : *yellow*
d : *dark green*
b : *black*
t : *tan* (brown)
B : *blue*
O : *orange*

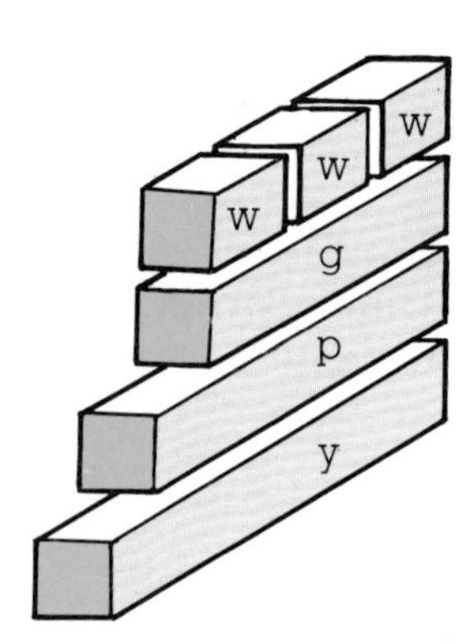

Question 1

Arrange the rods as shown in this diagram.

d					
g			*g*		
r		*r*		*r*	
w	*w*	*w*	*w*	*w*	*w*

Use them to solve the following problems.

1 $2g = \square$
2 $3\square = d$
3 $\square w = d$
4 $\square w = r$
5 $\square w = 2r$
6 $g = \square w$
7 $\square = \frac{1}{2}d$
8 $\square = \frac{1}{2}r$
9 $w = \square 2r$
10 $g = \square d$
11 $r = \square d$
12 $w = \square r$
13 $\square = (\frac{1}{3})g$
14 $2w = \square g$
15 $\square = (\frac{2}{3})d$
16 $\square = (\frac{4}{6})d$

Question 2

Write down any more fraction statements you can find using the same arrangement of rods.

Question 3

Now do the same for the following rod-picture.

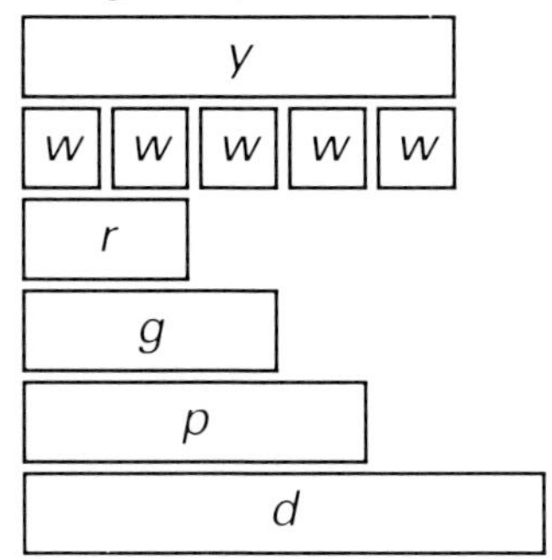

Find at least 20 fraction statements.

Question 4

Use sketches where possible to illustrate your answers to the following.
Give answer in terms of a single rod.

1 $(\frac{1}{3})g = \square$
2 $(\frac{2}{3})g = \square$
3 $(\frac{1}{5})y = \square$
4 $(\frac{1}{4})t = \square$
5 $(\frac{2}{5})y = \square$
6 $(\frac{3}{5})y = \square$
7 $(\frac{4}{5})y = \square$
8 $(\frac{5}{5})y = \square$
9 $(\frac{6}{5})y = \square$
10 $(\frac{3}{4})t = \square$
11 $(\frac{4}{3})g = \square$
12 $(\frac{5}{4})p = \square$
13 $(\frac{1}{3})(O+r) = \square$
14 $(\frac{4}{5})\ (O+y) = \square$
15 $(\frac{6}{7})\ (2b) = \square$

Now make up some more of your own.

Question 5

1 $\frac{1}{2} \times (\frac{1}{2} \times p) = \square$
2 $(\frac{1}{4})p = \square$
3 $\frac{1}{2} \times (\frac{1}{3} \times d) = \square$
4 $\frac{1}{3} \times (\frac{1}{3} \times B) = \square$
5 $(\frac{1}{9})B = \square$
6 $\frac{1}{2} \times (\frac{2}{3} \times d) = \square$
7 $(\frac{1}{3})d = \square$
8 $\frac{2}{3} \times (\frac{2}{3} \times B) = \square$
9 $(\frac{4}{9})B = \square$
10 $\frac{3}{4} \times (\frac{1}{2} \times (O+d)) = \square$
11 $\frac{3}{8}\ (O+d) = \square$
12 $\frac{1}{3} \times (\frac{4}{5}\ (O+y)) = \square$
13 $\frac{4}{15}\ (O+y) = \square$
14 $\frac{1}{3} \times (\frac{1}{4} \times (\frac{2}{3} \times (O+t))) = \square$
15 $\frac{1}{2} \times (\frac{1}{2} \times (\frac{1}{2} \times (\frac{1}{2} \times (O+d)))) = \square$
16 $\frac{2}{3} \times (\frac{1}{2} \times (\frac{3}{4} \times (\frac{1}{3} \times (4d)))) = \square$
17 $\frac{1}{18}\ (O+t) = \square$
18 $\frac{1}{16}\ (O+d) = \square$
19 $\frac{1}{12}\ (4d) = \square$
20 $\frac{3}{4} \times \frac{2}{3} \times \frac{3}{7} \times \frac{4}{5} \times (4t+g) = \square$

Question 6

Answer the following in terms of a single rod.

1 $\frac{1}{2} \times (\frac{1}{2} \times \square) = w$
2 $\frac{3}{5} \times (\frac{1}{2} \times \square) = g$
3 $\frac{1}{2} \times (\frac{4}{7} \times \square) = r$
4 $\frac{3}{4} \times (\frac{4}{5} \times \square) = d$

Fraction statements

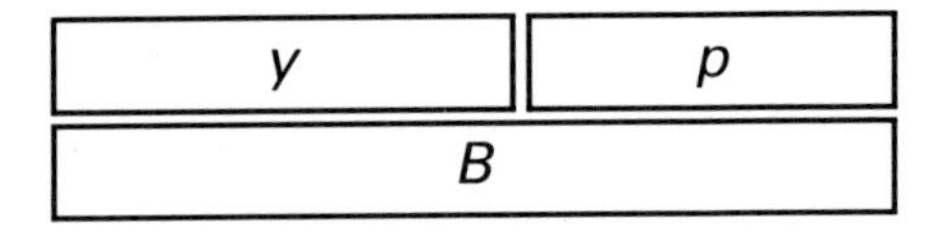

We know that $y + p = B$. If we let $w = 1$, then $y = 5$, $p = 4$ and $B = 5 + 4 = 9$. If $r = 1$, then $y = 2\frac{1}{2}$, $p = 2$ and therefore $B = 4\frac{1}{2}$. Work through the table and make sure you understand it all. You may need to refresh your memory of the different rod lengths by looking at the table at the beginning of the chapter. Then try question 7.

rod value	*statement*
$w = 1$	$5 + 4 = 9$
$w = 2$	$10 + 8 = 18$
$r = 1$	$2\frac{1}{2} + 2 = 4\frac{1}{2}$
$g = 1$	$1\frac{2}{3} + 1\frac{1}{3} = 3$
$p = 1$	$1\frac{1}{4} + 1 = 2\frac{1}{4}$
$y = 1$	$1 + \frac{4}{5} = 1\frac{4}{5}$
etc.	

Question 7

Do the fraction statements for the following two arrangements of rods:

(i)

g r
y

rod value	*statement*
$w = 1$	
$w = 2$	
$w = 3$	
$r = 1$	
$g = 1$	
$p = 1$	
$y = 1$	

(ii)

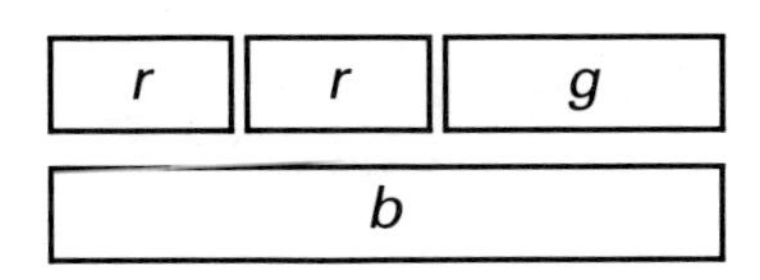

rod value	*statement*
$w = 1$	
$r = 2$	
$g = 3$	
$p = 1$	
$y = 1$	
$d = 1$	
$b = 1$	
$r = ½$	

Now draw a rod arrangement of your own and make up some rod values like those in the table above. Then fill in the statements or swap your problems with your friends.

Question 8

It is true that

$$t = 2p = 4r = 8w$$

which we can set up as:

t

p p

r r r r

w w w w w w w w

We can write lots of statements about this picture such as:

$$2r = p \quad 6w = 3r \quad 4w = p$$

Let us make the rod value $t = 1$.

(i) What are the values of p, r and w?

(ii) Explain why we can now say that $4w = p$ is the same as

$$4 \times \frac{1}{8} = \frac{1}{2} \text{ (a fraction statement)}$$

(iii) Make up as many fraction statements based on the diagram as you can.

13 Manhattan Police

The streets in New York form a square grid. The buildings in each section are called **blocks**.

Imagine that a policeman in Manhattan can only see as far as the next block. To see as far as possible, he must stand on a corner of a street.

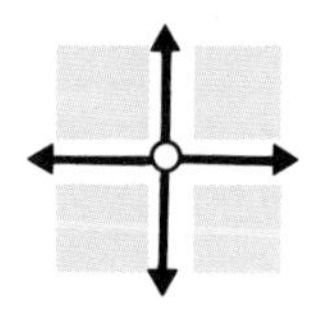

This policeman can cover 4 blocks.

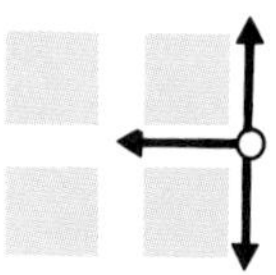

This one can only cover 3 blocks.

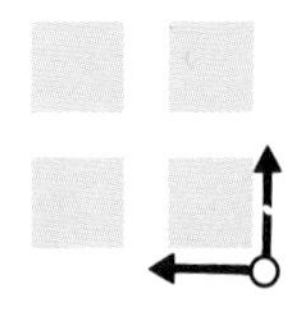

This one only manages 2 blocks.

In questions 1 and 2 all the blocks are in a row.

Question 1

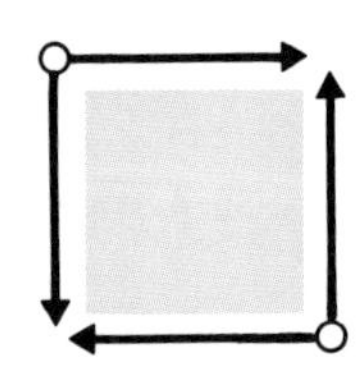

If we have 1 block then we need 2 policemen to keep watch.

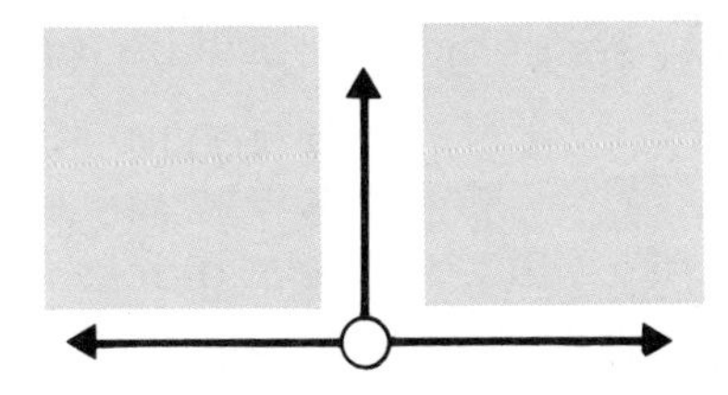

This time we have 2 blocks to cover. 1 policeman is already in place for his watch.

How many more do we need to cover all sides of the blocks? Try not to get 2 policemen looking down the same street.

Continue your investigation further for

1 3 blocks
2 4 blocks
3 5 blocks in a row

and so on.

Can you predict how many policemen you would need for

4 10 blocks
5 20 blocks
6 30 blocks
7 100 blocks
8 1000 blocks
9 375 blocks
10 n blocks in a row?

It may help if you set out your results in a table like this:

blocks in a row	*policemen*
1	2
2	
3	
⋮	

Question 2

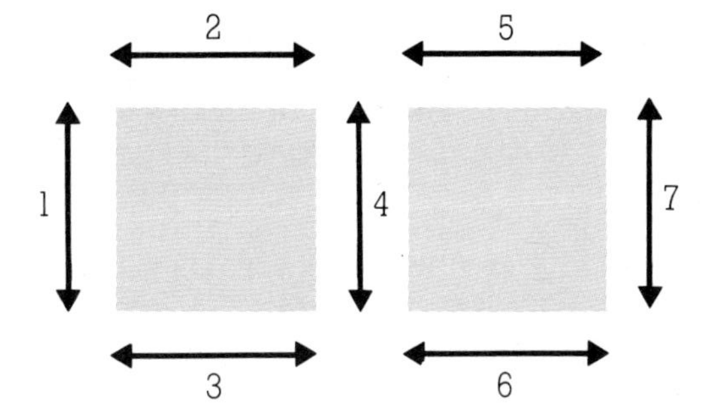

If we look at these two blocks, then we see that we have 7 sections to watch over.

Can you predict how many sections there will be for

1 5 blocks
2 10 blocks
3 40 blocks
4 100 blocks
5 737 blocks
6 *n* blocks in a row?

It may help you to set out your results like this:

blocks in a row	*sections*
1	
2 →	7
3	
4	
5	
6	
⋮	

Question 3

1×1 block 2×2 block 3×3 block

In this question the blocks form a square grid. Investigate the number of policemen needed in each case. Note that there are more ways than one of finding how many police are needed. You must find the least number. Set your results out like this:

grid size	*number of blocks in grid*	*policemen*
1		
2		
3		
4		
5		
⋮		
n		

Question 4

Find the least number of policeman needed to patrol these blocks.

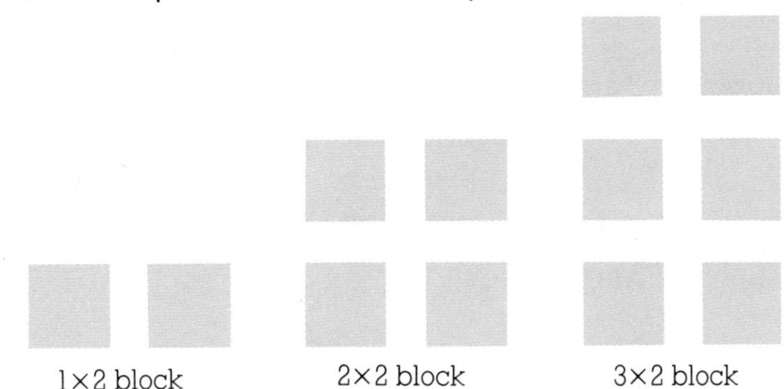

(i) Investigate this pattern very carefully as it is not very easy to follow. Can you predict for a $n \times 2$ block?

(ii) Investigate *at least* one more pattern. For example

$1\times3, 2\times3, 3\times3, 4\times3$, etc.
or $1\times4, 2\times4, 3\times4, 4\times4$, etc.

You will find that 1×4 is the same as 4×1 which will save you time.

Question 5

An area of 5 blocks can be arranged in a number of different ways. For example this arrangement uses 6 policemen.

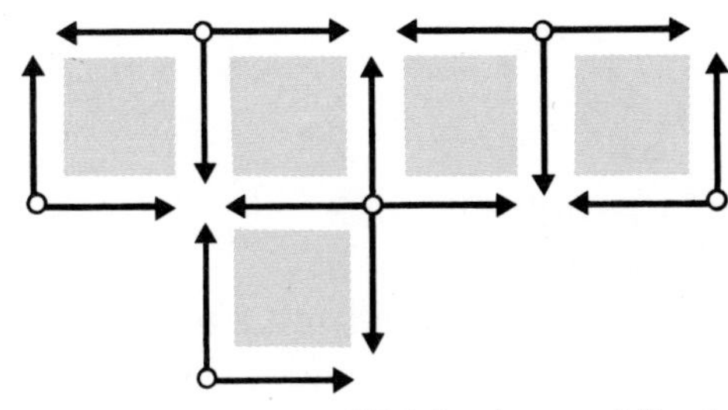

(i) Investigate the arrangements of 5 blocks and find the one which uses *least* police.

(ii) Investigate in a similar way for

1 4 blocks
2 3 blocks
3 6 blocks
4 7 blocks
5 8 blocks.

(iii) What can you say about the most *efficient* grouping of the blocks?

14 Number Theory 1
3's and 5's

Some **multiples** of 3 are:

$$18 \quad \text{written as} \quad 6 \times 3 \text{ or } 6(3)$$
$$30 \longrightarrow 10(3)$$
$$9 \longrightarrow 3(3)$$

Some multiples of 5 are:

$$15 \longrightarrow 3(5)$$
$$30 \longrightarrow 6(5)$$
$$100 \longrightarrow 20(5)$$

Some numbers are multiples of both 3 *and also* of 5, for instance:

$$30 = 10(3) = 6(5)$$

Some numbers can be made up by *adding* a multiple of 3 to a multiple of 5, for instance:

$$17 = 12 + 5$$
$$= 4(3) + 1(5).$$

However some cannot, e.g. 4.
Note: for this work the use of either fractions and/or negative signs is strictly against the rules. However zero can be used.

Question 1

Write down:

1 four numbers which are multiples of 3
2 four numbers which are multiples of 5
3 two numbers which are multiples of both 3 and 5.

Question 2

Express each of the following as a multiple of 3 plus a multiple of 5.

1 8	**3** 47
2 19	**4** 80

If you believe it cannot be done, then explain why.

Question 3

Express each of the following numbers as a multiple of 3 plus a multiple of 5 in as many ways as you possibly can:

1 29
2 38
3 67.

Make any observations you can.

Question 4

Where possible express each of the numbers 1, 2, 3, 4, . . . as a multiple of 3 plus a multiple of 5. Some have been done. A star means that it cannot be done.

1 = ★
2 = ★
3 = 1(3) + 0(5)
4 = ★
5 = 0(3) + 1(5)
⋮

Do not go on for ever. Stop when you have found the number which you believe is the largest that *cannot* be made by adding a multiple of 3 to a multiple of 5.
Explain why you think this is the case.

Question 5

Instead of using 3 and 5, repeat the previous question using

1 4 and 7
2 5 and 8
3 2 and 9
4 5 and 9
5 3 and 7.

Can you give a **general** result? Try a few more examples if you need to.

Question 6

Try the same exercise using

1 3 and 6
2 4 and 6
3 6 and 15
4 8 and 12.

Do the results here change the general result at all?

Question 7

Given any two positive integers, what is the most general statement you can make about the largest number that *cannot* be made by adding positive integer multiples of the two given integers?

Question 8

Prove or explain in your own words, the general result obtained in the previous question. This will be difficult so do not feel discouraged if you cannot do it.

Question 9

If the cost of 1st and 2nd class postage stamps are 18p and 13p respectively, show that it is possible to make all postage costs over £2.03 using combinations of these stamps.

15 Number Bases 1

Converting from one base to another

Grouping in circles

Here is a method for converting numbers from one base to another. It is called **grouping**. The rules are:

(i) You can have as many circles as you need but you may only have 15 counters.

(ii) The unit circle where you start is on the *right*.

(iii) The base you are changing *to*, determines the size of the group of counters worth 1 in the next circle along.

Example

Converting 12 in base 10 to base 3

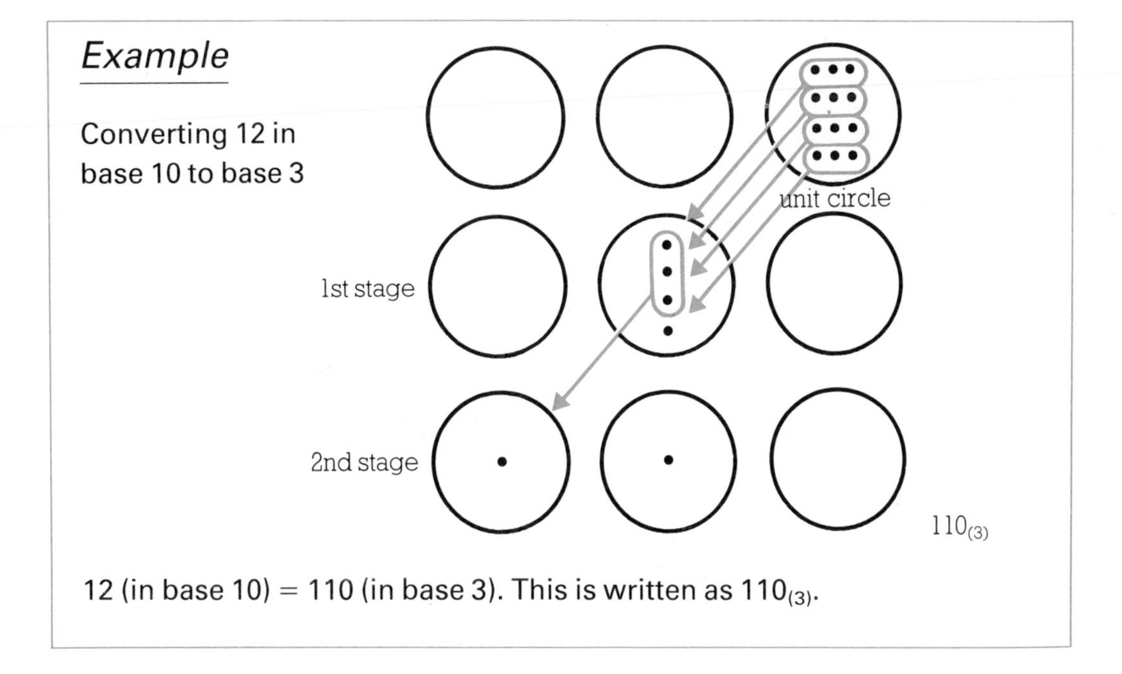

12 (in base 10) = 110 (in base 3). This is written as $110_{(3)}$.

Question 1

Use the method of **grouping** to convert the following base 10 numbers to the bases shown on the next page.

base 10	*base* 2	*base* 3	*base* 5	*base* 8
3				
7				
9				
11				
16				
21				
25				
27				
31				

Try some more of your own.

Explain how you have coped in the cases where the numbers are larger than 15 (the number of counters you are allowed).

Converting back to base 10

Doing the reverse procedure, you move counters back towards the unit circle. As each counter moves back a circle, its worth becomes that of the base value. The base 10 equivalent is given by the number of counters in the right-hand circle when the process is complete.

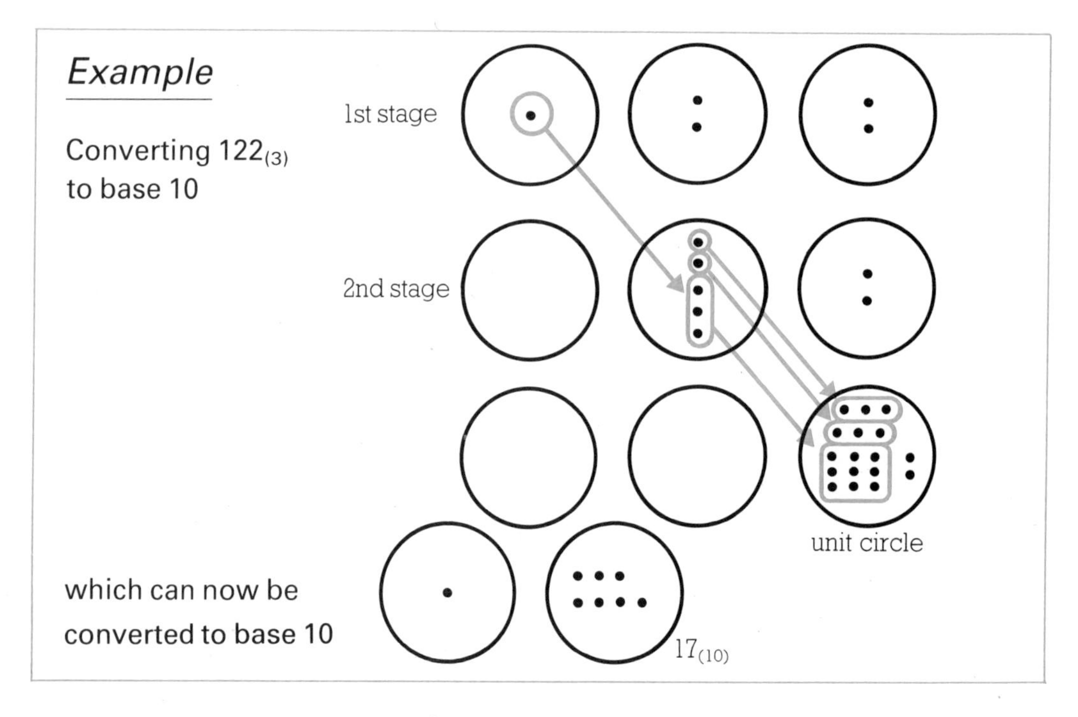

Question 2

Complete the following table using the methods of converting shown in this chapter.

base 10	*base* 2	*base* 3	*base* 5
	10011		
	111011		
		1020	
			43

Now try some more of your own.

Quick methods

For converting from base 2 to base 10:

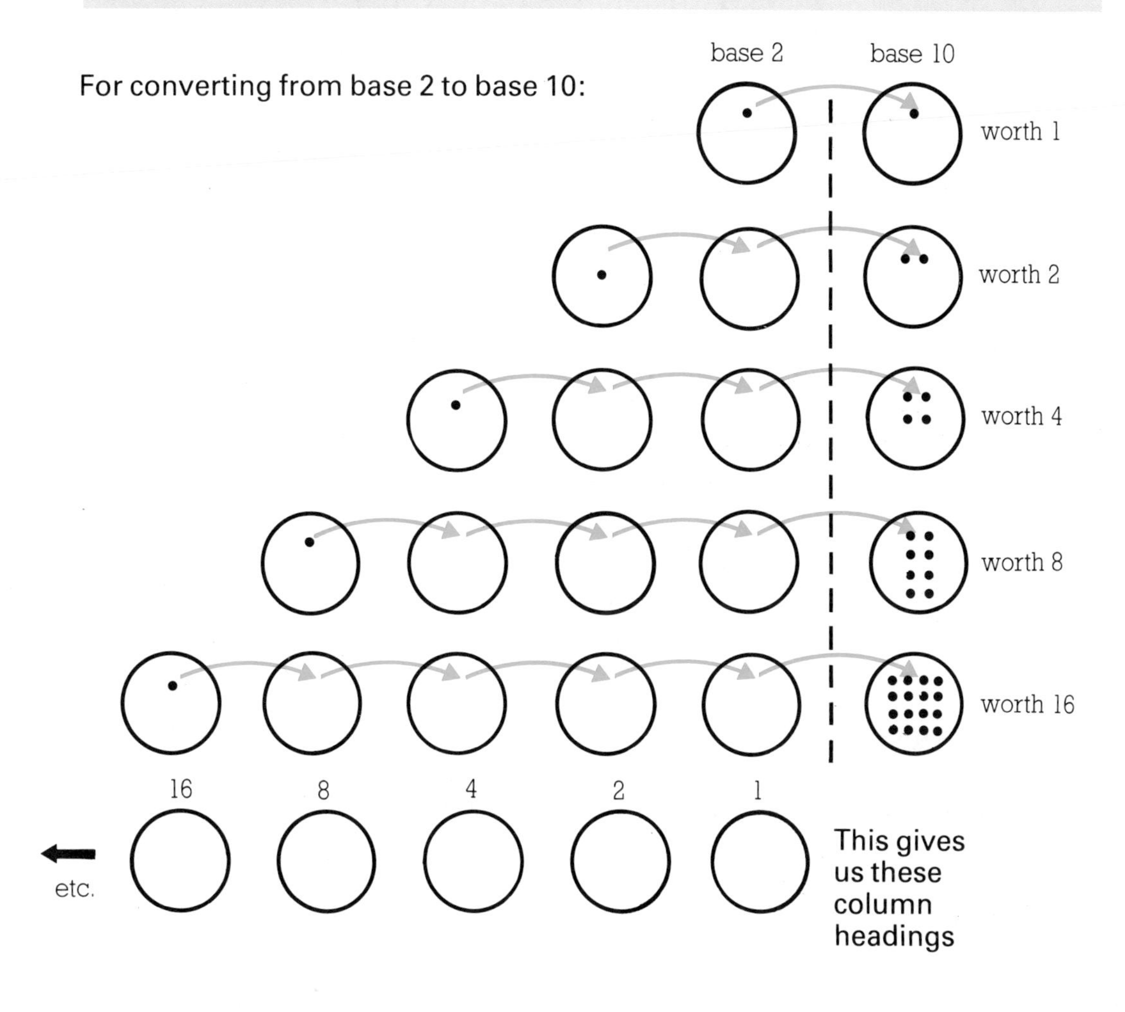

Therefore $1101101_{(2)}$ can be quickly converted to base 10 as follows:

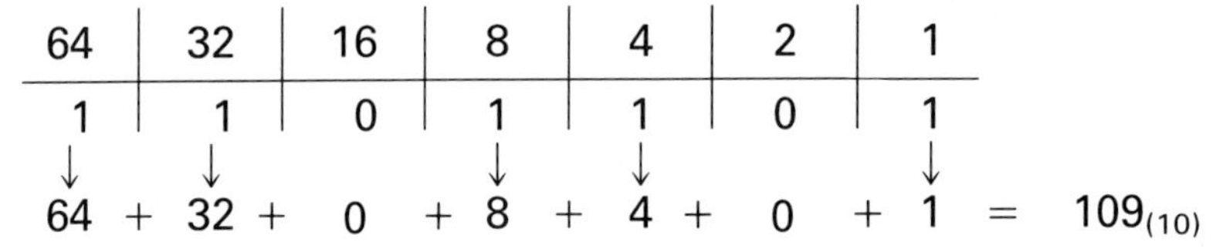

64	32	16	8	4	2	1
1	1	0	1	1	0	1

$64 + 32 + 0 + 8 + 4 + 0 + 1 = 109_{(10)}$

Similarly for base 3:

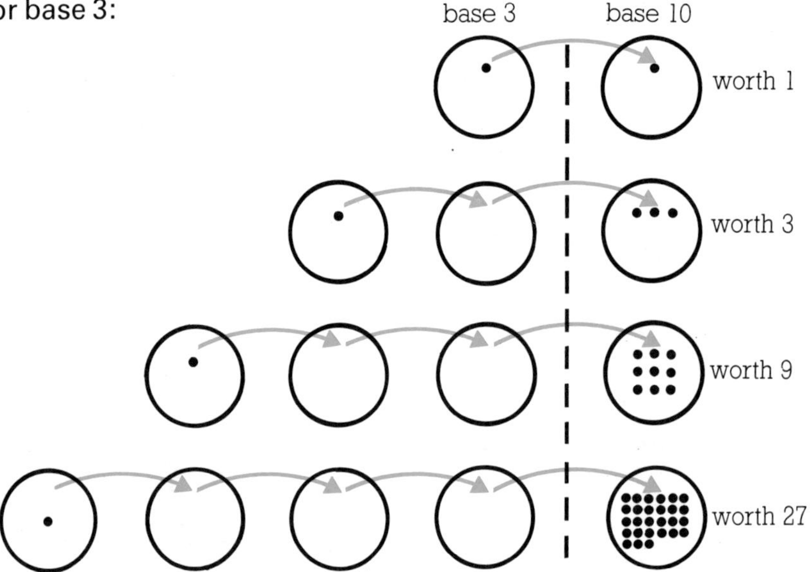

This gives column headings of:

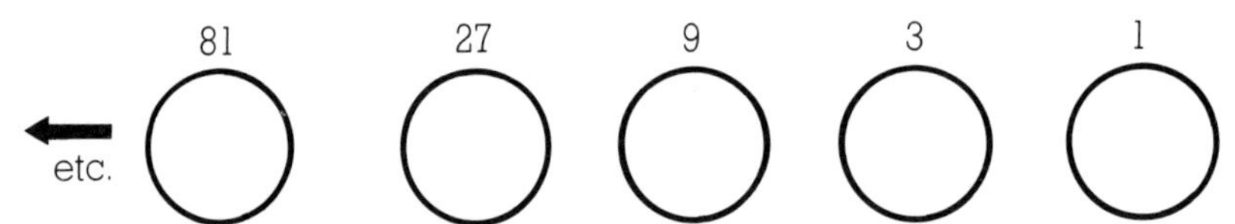

So that $12011_{(3)}$ can be quickly converted as follows:

81	27	9	3	1
1	2	0	1	1

$81 + 54 + 0 + 3 + 1 = 139_{(10)}$

Question 3

Use the quick methods to convert the following numbers back to base 10:

1 $110110_{(2)}$
2 $110011011_{(2)}$
3 $110112_{(3)}$
4 $202020_{(3)}$
5 $10101_{(4)}$
6 $301403_{(5)}$.

Try some more of your own.

Repeated division

For question 4, either use the quick method or use the method shown below. Can you see the significance of the remainders (R) when there is repeated division by the base?

Example

Convert $37_{(10)}$ to base 5: $5\,\underline{|\,37}$
$5\,\underline{|\,7}$ R **2**
1 R **2**

Answer: $37_{(10)} = 122_{(5)}$

Similarly,
Convert $27_{(10)}$ to base 4: $4\,\underline{|\,27}$
$4\,\underline{|\,6}$ R **3**
1 R **2**

Answer: $27_{(10)} = 123_{(4)}$

Question 4

Fill in the positions marked by stars in the following table:

base 10	*base 2*	*base 3*	*base 5*	*base 8*
71	★		★	
128		★		★
209		★	★	
671			★	★

Try some more of your own.

16 Geoboard 2
Finding areas

Doubling and halving

We can find the areas of shapes in a variety of ways.

We will take this square to represent one unit of area.

One method you will have discussed is **doubling** and **halving**.

Example

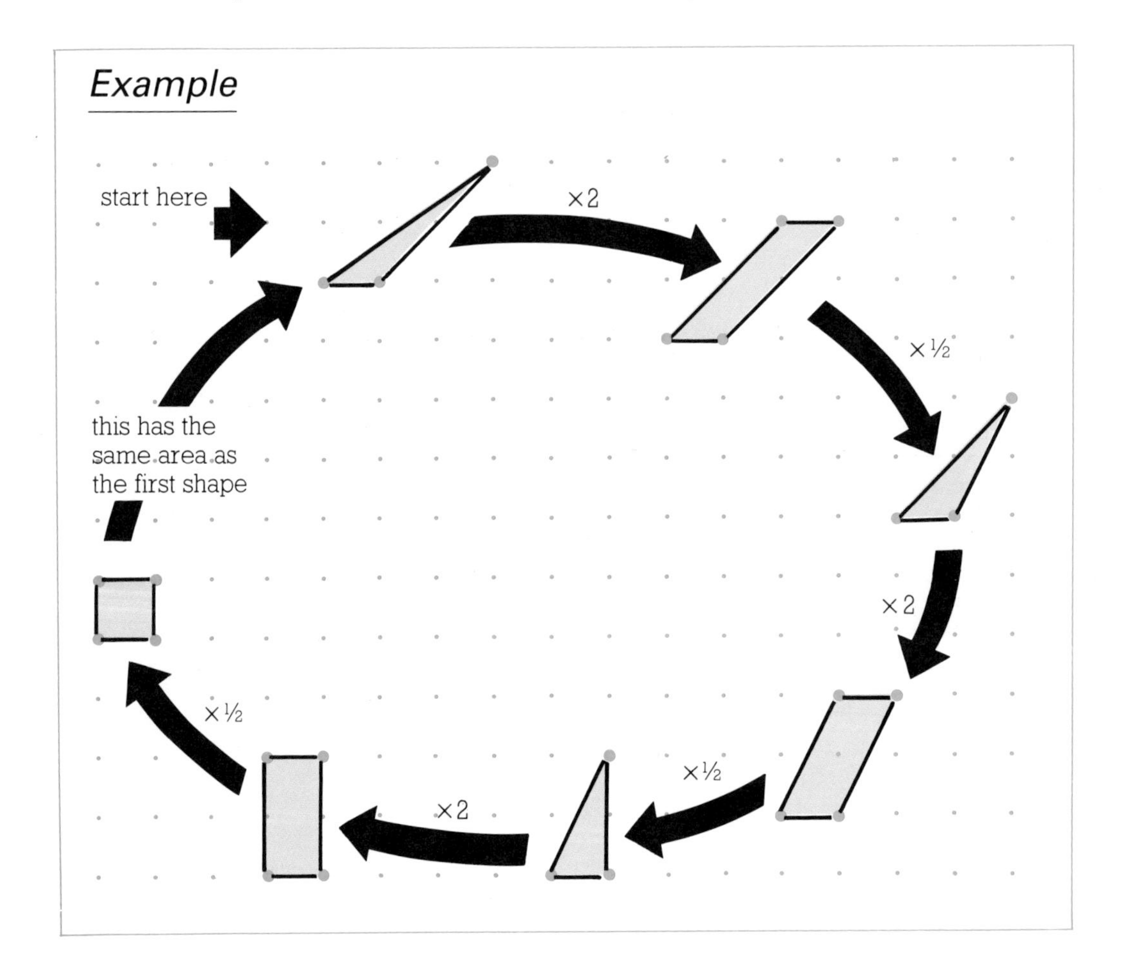

Question 1

Use the method in the example to find the areas of the following shapes:

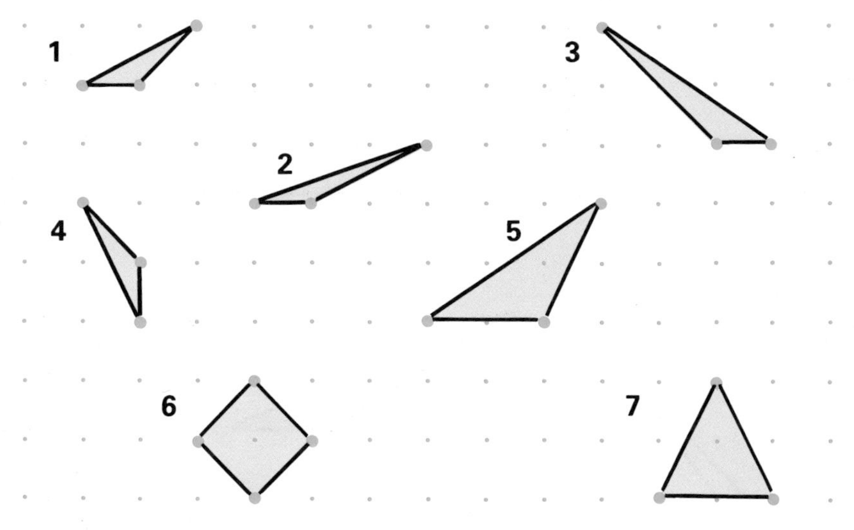

Another method, when finding the areas of triangles, is applying the formula:

$$\text{area} = \frac{1}{2} \text{ of (base)} \times \text{(perpendicular height)}$$

Example

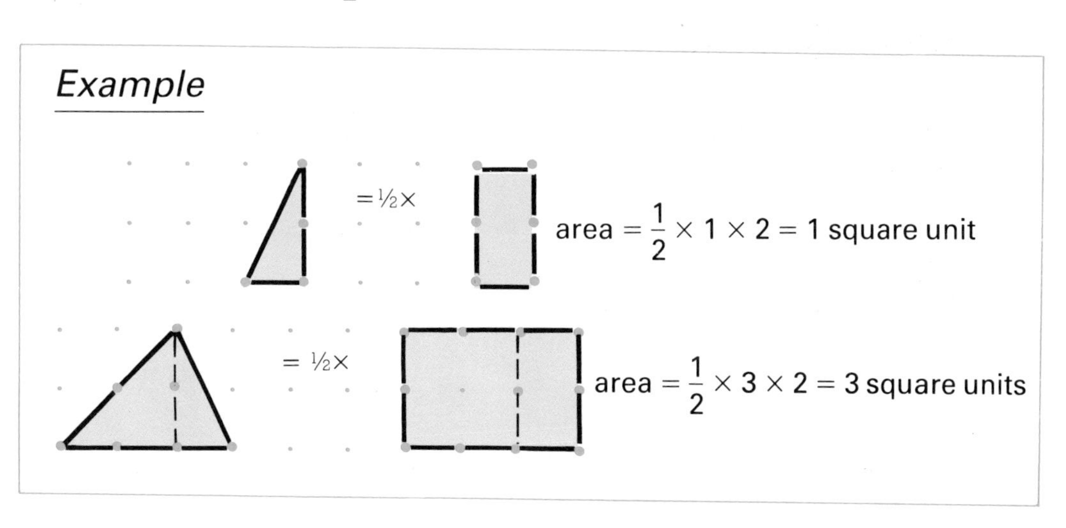

Question 2

Now use the formula (where possible) to find the areas of the following:

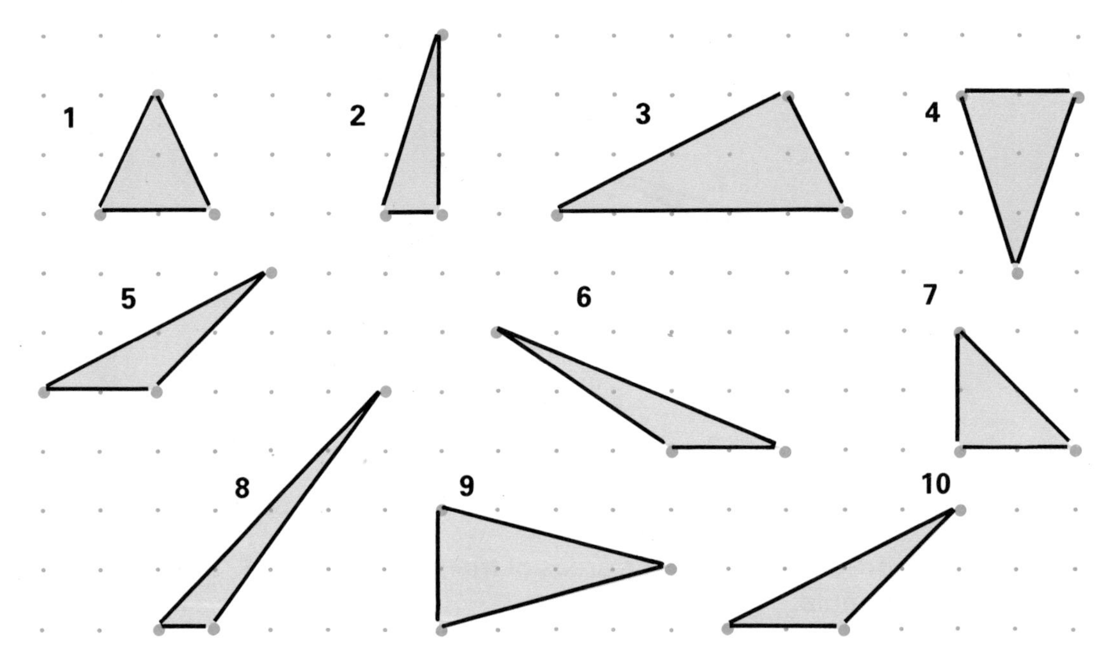

Internal dissection

This method can be used when it is easier to break-up a shape into simple components.

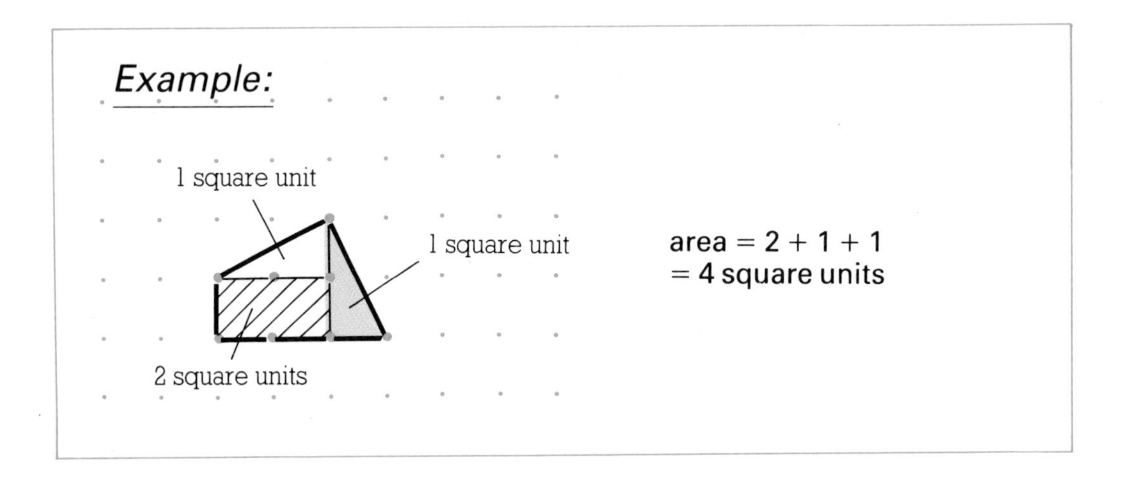

Question 3

Use this method to evaluate the areas of the following shapes:

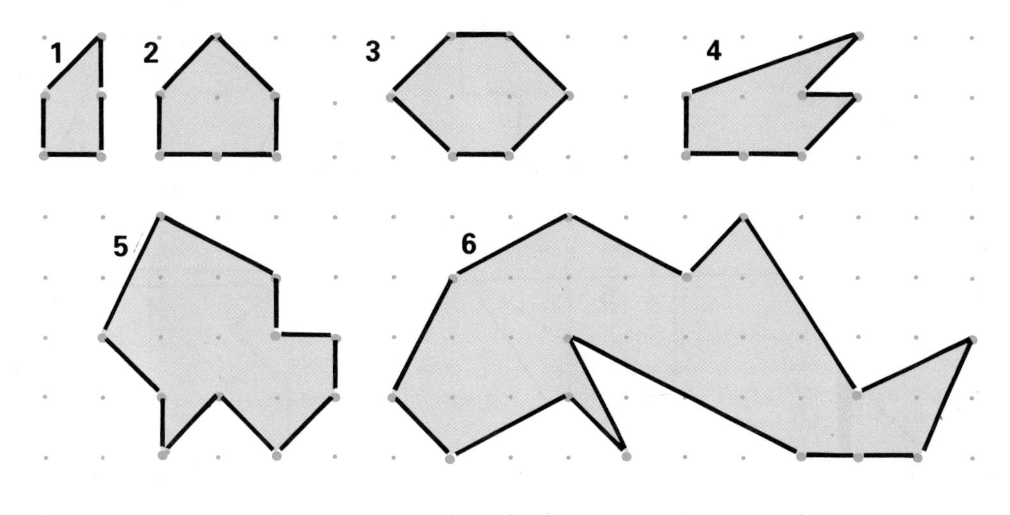

External dissection

This can be used when it is easier to *subtract* areas from a surrounding rectangle.

Example

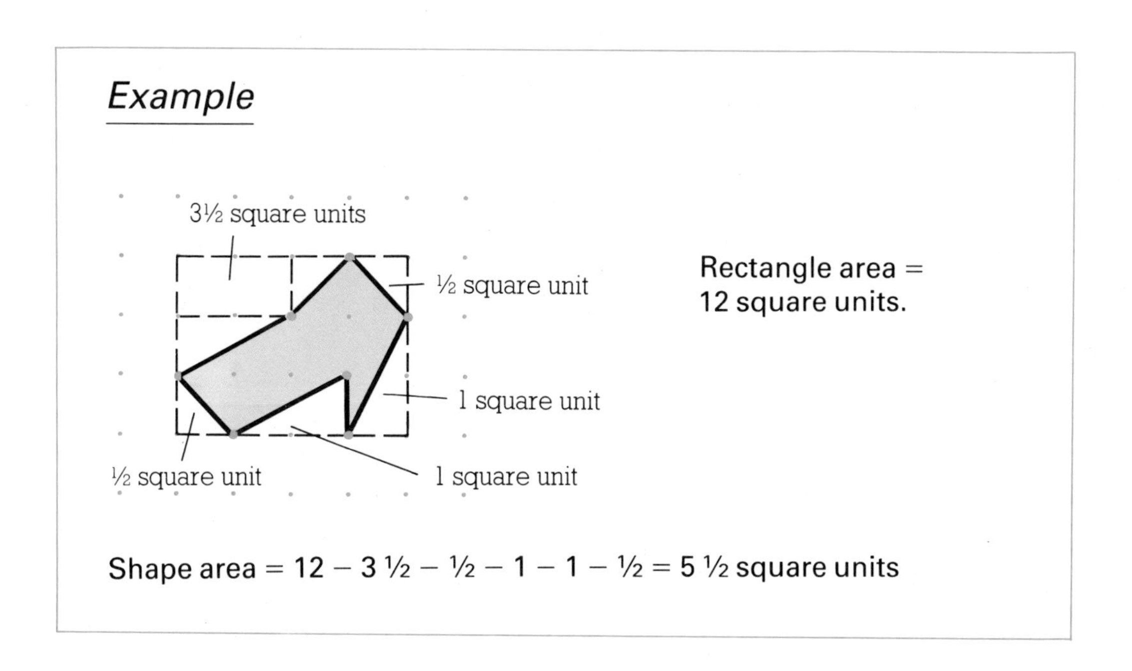

Rectangle area = 12 square units.

Shape area = 12 − 3 ½ − ½ − 1 − 1 − ½ = 5 ½ square units

Question 4

Use this method to evaluate the following areas. Show your constructions.

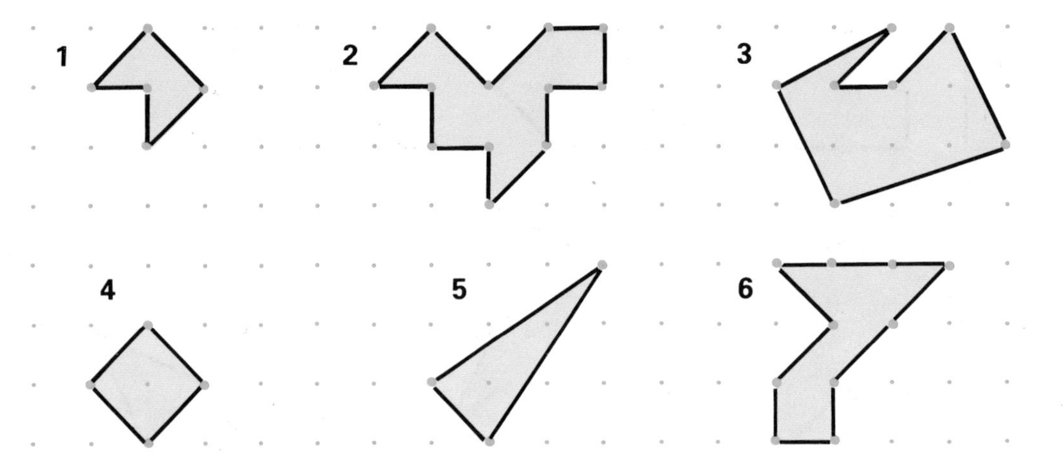

Question 5

Find the areas of the following shapes, explaining clearly why the method you have chosen is the most appropriate in each case:

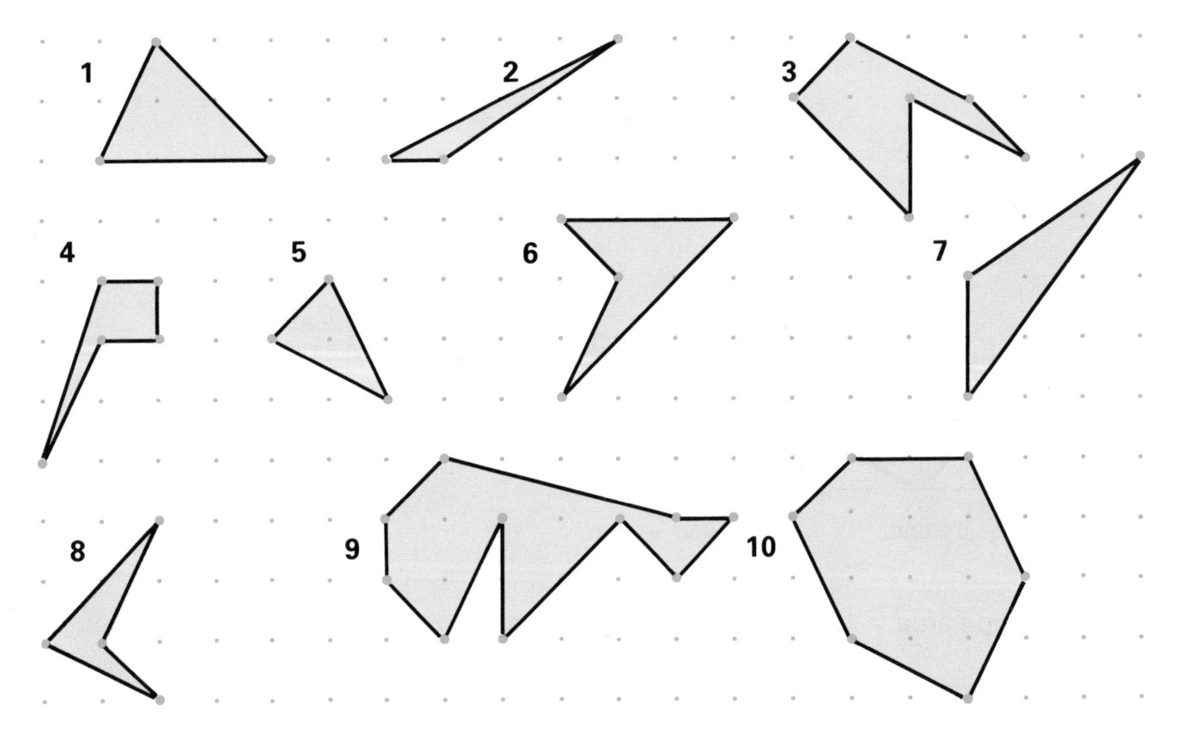

Special questions

Question 6

Can you show clearly why the area of the shaded triangle (ADE) equals $2/7$ of the 16-pin square, ABCD?

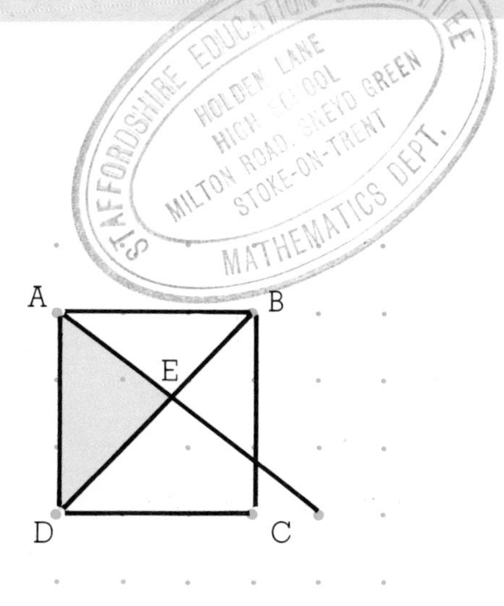

Question 7

Find what fraction of the board the shaded areas represent in the following cases:

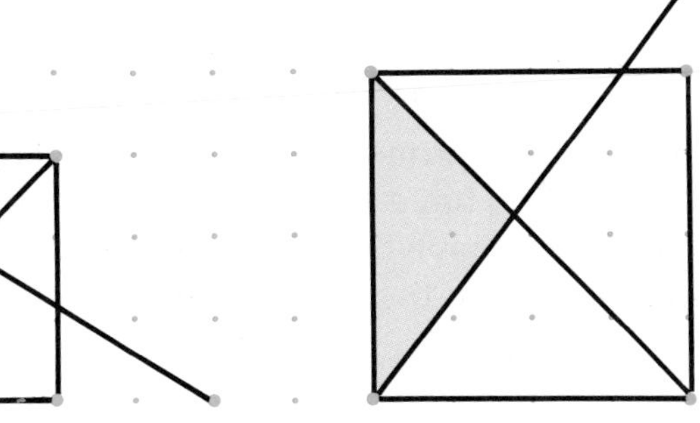

Question 8

Find the area of the shaded overlap region in the problem shown: where C is the centre of the smaller of the 2 squares shown, and the side AB has been trisected.

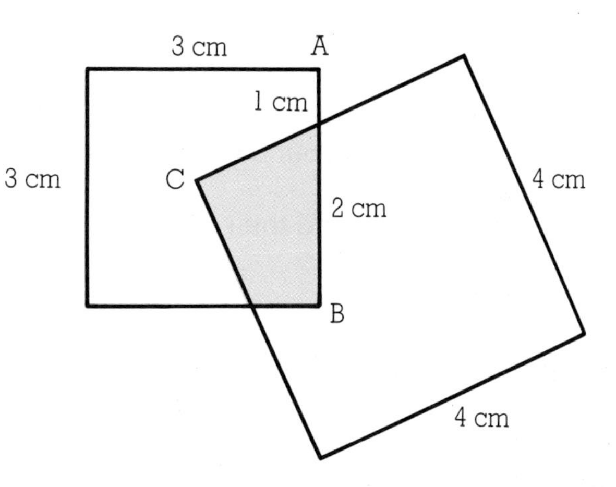

17 Crossovers

The diagonal of a 3×5 grid has been drawn. It makes 6 **crossovers** with the grid lines.

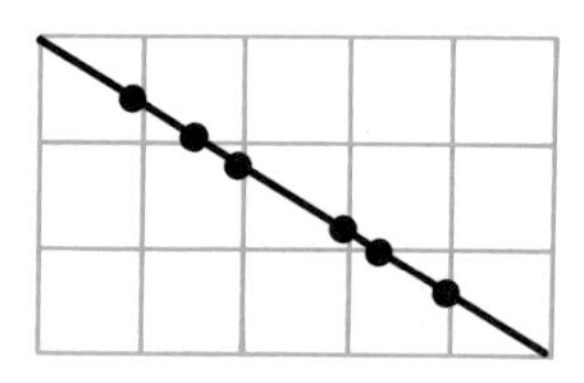

Your task is to look at the relationship between the number of crossovers and the grid sizes.

Question 1

Draw grids of sizes: 3×7 4×5 5×8
7×4 6×11 7×5

Draw in the diagonals and count and record the numbers of crossovers.

Perhaps you have already seen a relationship. A common feature of all the grid sizes you have been given so far is that each pair of numbers is **co-prime**; that is, their only common factor is 1.

Question 2

These grids have pairs of numbers which are not co-prime:

4×6 2×4 3×9
6×15 8×10 9×12

Draw each grid. Draw in the diagonal on each. Count and record the numbers of crossovers. Does your earlier relationship break down? If so why does it break down?

Draw more grids and their diagonals to check your conclusions.

Question 3

Suppose you had grids of sizes:

20×17 and 72×30

Can you work out the number of crossovers the diagonal makes? Can you **generalise** the result?

18 Cuisenaire Activity 3
Rod products and permutations

Rod values

w : *white*
r : *red*
g : *light green*
p : *pink*
y : *yellow*
d : *dark green*
b : *black*
t : *tan* (brown)
B : *blue*
O : *orange*

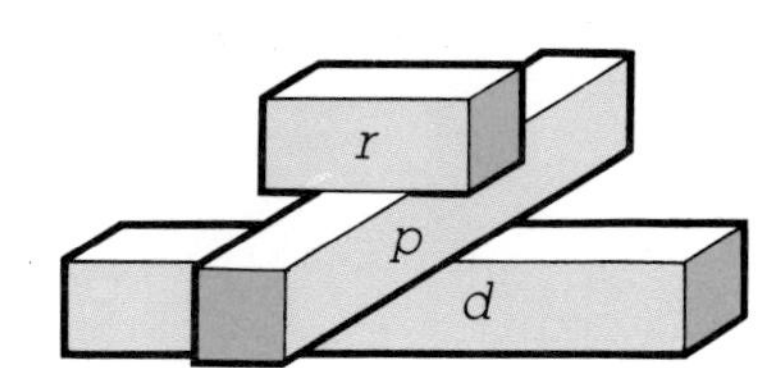

When different colour rods are used as a base, then colour values can differ so that when

$w = 1$, then $y = 5$

y				
w	*w*	*w*	*w*	*w*

because it is 5 times longer than *w*. But when

$w = 2, y = 10$

When

$y = 1$ then $O = 2$ and $w = 1/5$

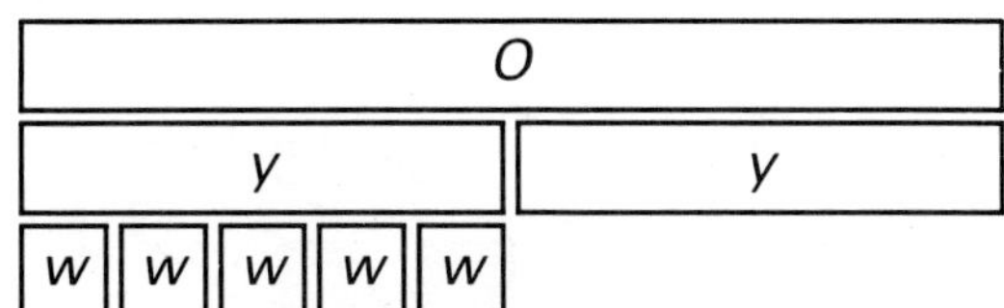

Use these principles for question 1.

Question 1

	base rod	find the value of
1	$w = 1$	O ; b ; B ; y ; $3r$
2	$w = 2$	O ; B ; p ; y ; $3p$
3	$w = 4\frac{1}{2}$	r ; p ; g ; O ; t
4	$r = 1$	O ; p ; w ; y ; g
5	$g = 1$	B ; d ; w ; r ; $5g$
6	$y = 1$	w ; p ; b
7	$b = 1$	w ; r ; g ; t
8	$p = 5$	t ; r ; w ; b

Rod products (building towers)

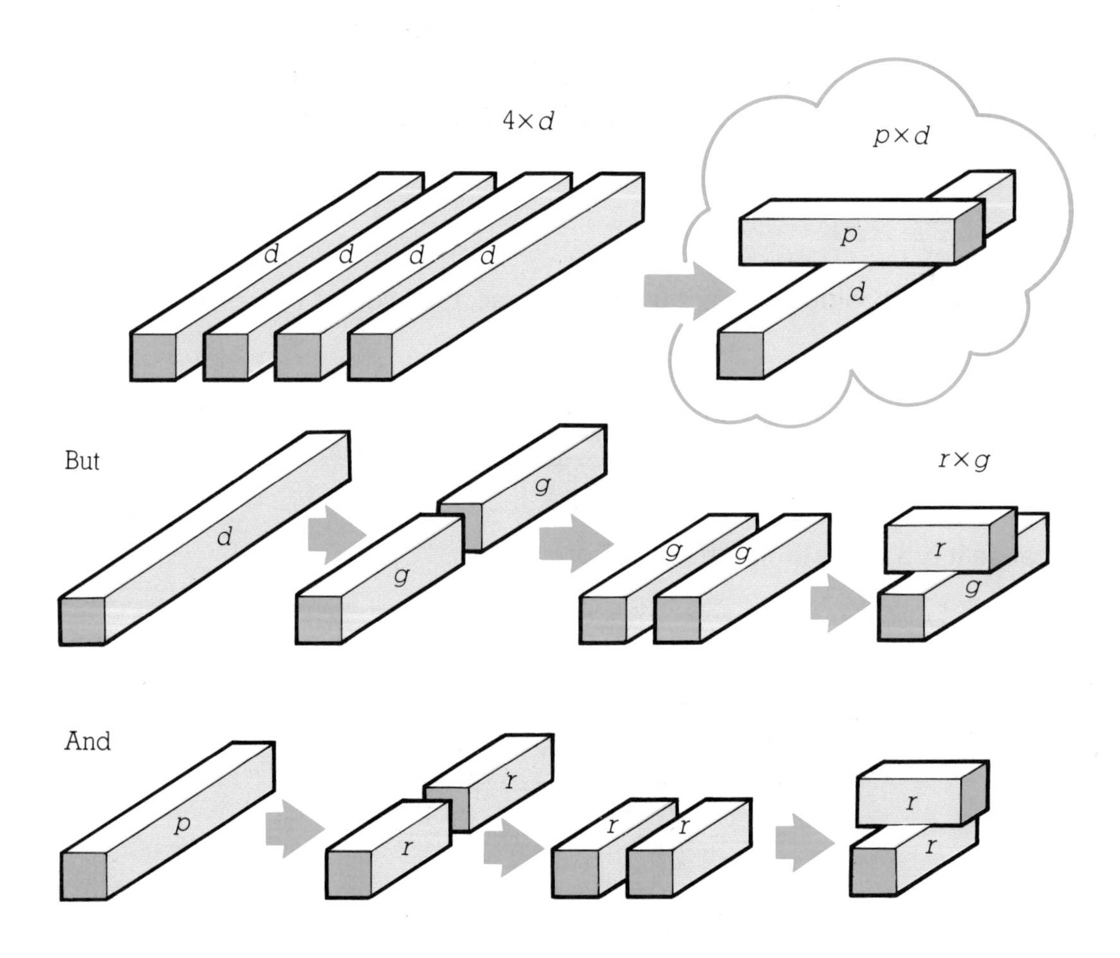

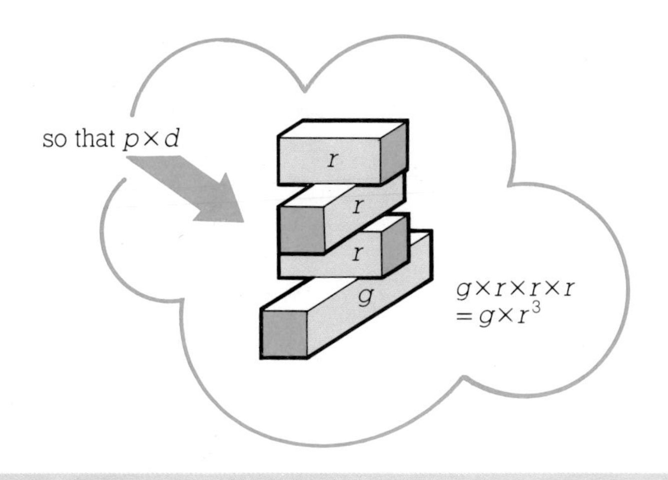

Equivalent forms

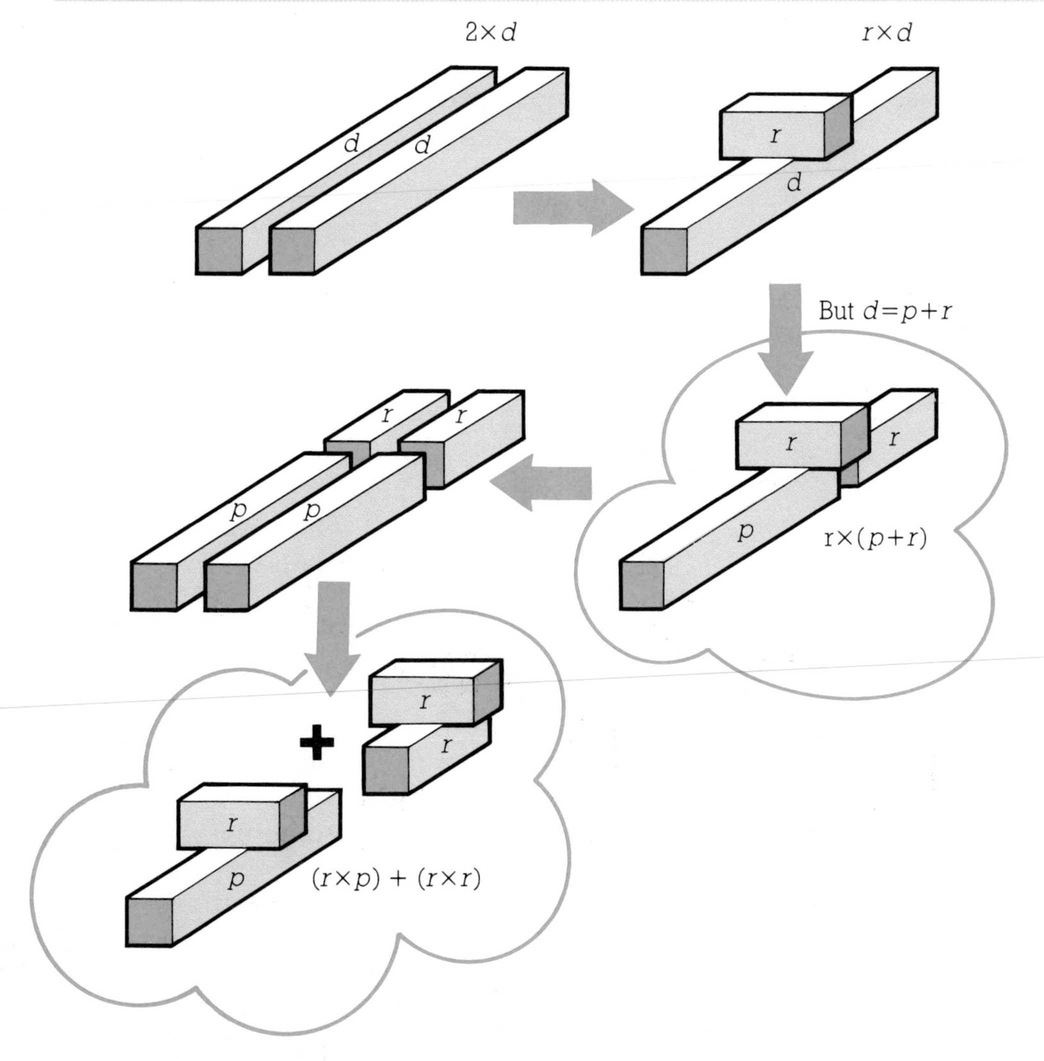

Question 2

Now use rod products and equivalent forms to test which of the following statements are true. Where a statement is wrong, try to give a correct version.

1 $p \times p = r \times r \times r \times r [= r^4]$
2 $y \times d = y \times (r \times g)$
3 $p \times p \times p = r \times r \times r \times r \times r [= r^5]$
4 $O \times r = r \times r \times g$
5 $d \times p \times d = r \times r \times r \times g \times g [= r^3 \times g^2]$
6 $(r \times g)^2 = r^2 \times g$
7 $(r \times p)^2 = r^6$
8 $(r \times g)^3 = r^3 \times g^3$
9 $(r^2)^3 = r^5$
10 $(r \times g \times B)^2 = p \times g^6$

Question 3

Fill in the missing powers in the following questions:

1 $(p \times d)^3 = r^{\boxed{?}} \times g^{\boxed{?}}$
2 $(g \times B)^2 = g^{\boxed{?}}$
3 $(t \times p) = r^{\boxed{?}}$
4 $(t \times p)^2 = r^{\boxed{?}}$
5 $(d \times p \times t \times B) = r^{\boxed{?}} \times g$

Permutations

Question 4

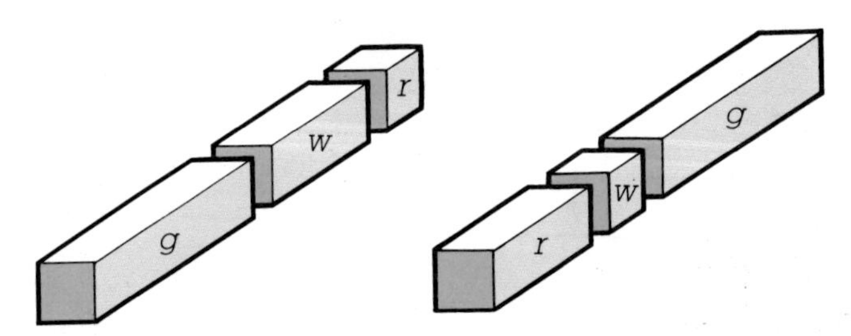

(i) When 3 different rods are laid end-to-end, there are 6 different ways this can be done.
If the ones above are *g r w* and *r w g* find the other **permutations**.
(ii) Use rods to fill in the table

number of different colour rods	*number of permutations*
1	
2	
3 →	6
4	
5	
etc.	

(iii) Can you explain and predict the results?

How many arrangements are there in the following questions where some of the rods are identical?

1 *w w r g*
2 *w w w r g*
3 *w w r r*
4 *w w r g p*
5 *w w r r g p*

Now try some of your own. Can you establish a method to predict the results?

Question 5

Suppose we have a rectangle which measures 2 × 3 cm. We could fill this with rods in different ways, i.e.

6 *whites*

w	w	w
w	w	w

4 *whites* 1 *red*

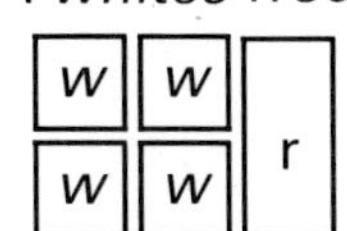

How many different ways could we fill the rectangle with rods? (Be careful; you must decide what you mean by *different*.)

19 Angles in Polygons

We can examine the angles in a triangle by moving an arrow around the sides of the triangle. The arrow *slides* along a side and *turns* through an angle. Every turn must be anticlockwise. Notice that you can move the arrow backwards as well as forwards.

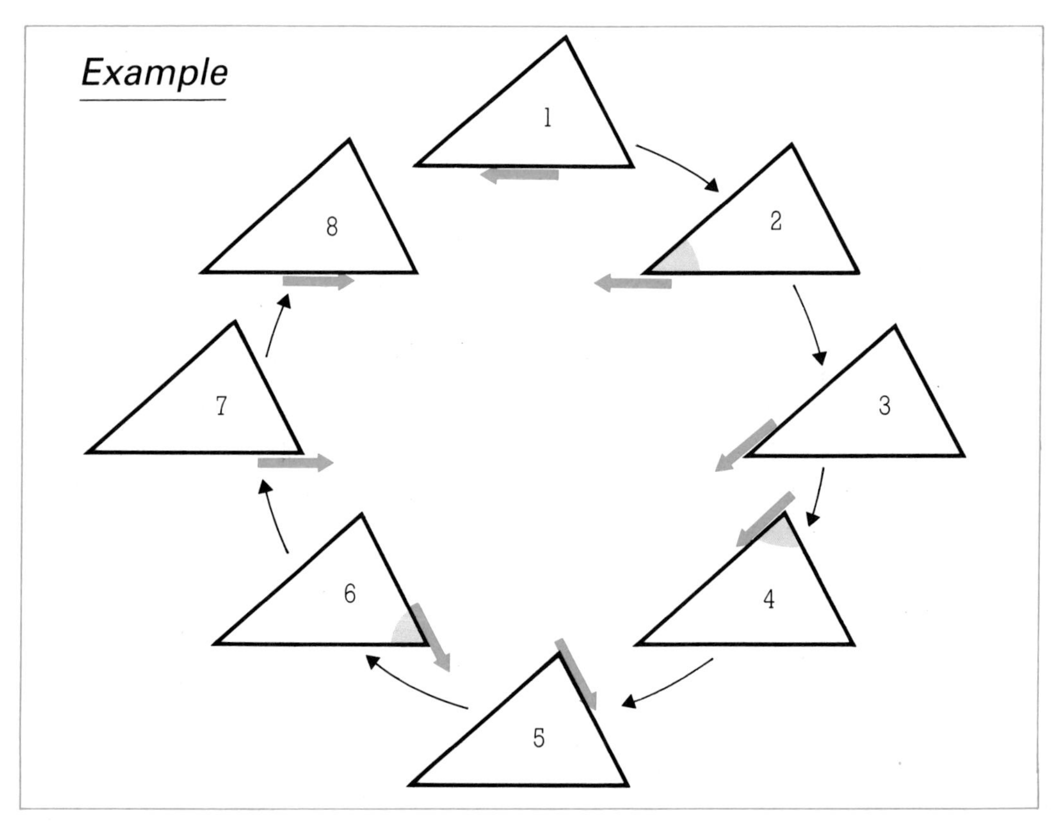

As the arrow moves completely around the triangle it goes from ⟵ to ⟶ which means it has turned through exactly **2 right angles** or 2 lots of 90° or 180°.

Question 1

Draw 3 triangles of your own. Move your arrow completely around each triangle. If you get it to turn through something *other* than 2 right angles in any case, show your teacher.

Question 2

Draw *three* 4 sided figures. Move your arrow completely around each one. How many right angles does it turn through in each case? Your answers should all be the same. If they are, go on to the next question. If they are different, speak with your teacher.

If it turns through 4 right angles in all cases, move on to the next questions.

Question 3

Do what you have done with the triangles and 4 sided figures for:

1 two 5 sided figures
2 two 6 sided figures
3 one 7 sided figure.

Count and record the number of right angles the arrow turns through in each case.

Question 4

Put your results into a table like the one below:

number of sides	*number of right angles turned through*	*degrees*
3	2	180
4		
5		
6		
7		

Extend the table for 8, 9, 10 sides without drawing any more figures if you can.

Question 5

Look at the number patterns in the table. Discuss them with your neighbours and teacher. Can you see any special pattern? Could you predict the number of right angles turned through for figures with:

1 20 sides
2 50 sides?

Question 6

Imagine that a figure had 1000 sides. Show all the working you would do to work out how many right angles the arrow would turn through.

Question 7

Could you give a **general formula** if the figure had *n* sides? You may need to discuss this with your teacher or neighbours.

Question 8

In each of the cases below, calculate the angle marked *x*. All the angles are in degrees.

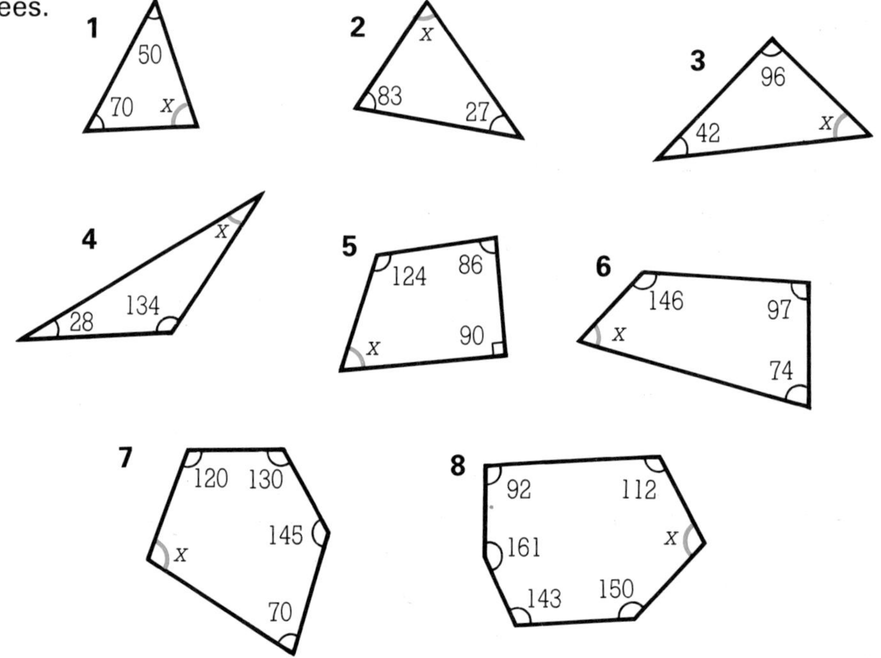

Question 9

Calculate the interior angles of:

1 an equilateral triangle
2 a square
3 a regular pentagon (5 sides)
4 a regular hexagon (6 sides)
5 a regular octagon (8 sides).

Question 10

The diagram shows one of the interior angles of a regular polygon.

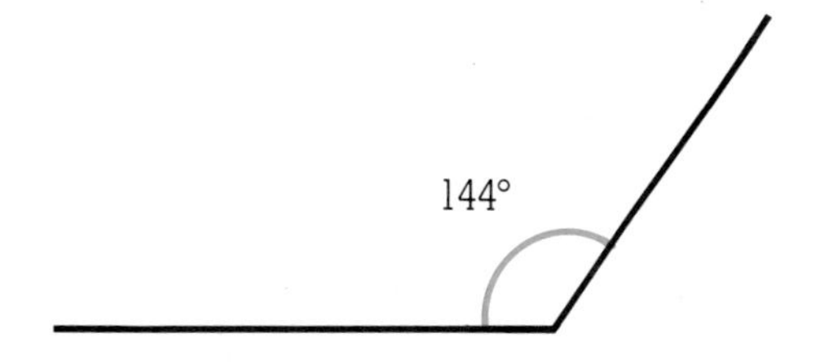

How many sides does the polygon have?

Question 11

Calculate the interior angles for regular polygons with:

1 10 sides
2 15 sides
3 20 sides.

Question 12

Use all the results you have obtained so far to draw the following graph:

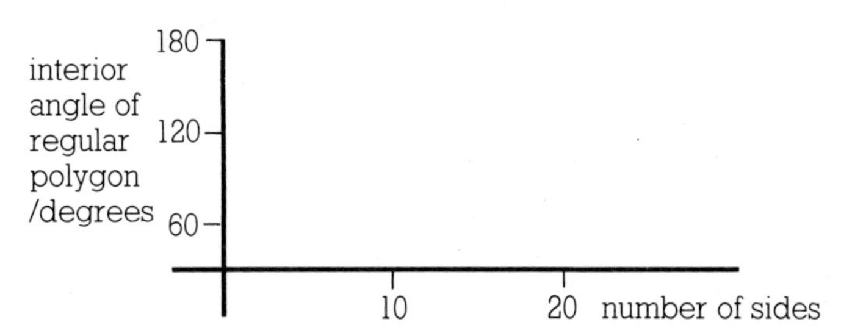

As the number of sides gets larger and larger, what do you think happens to the interior angle?

20 Hidden Dot Shapes

Your task in the following exercises is to find some hidden shapes. The only information you have is that the crosses are at the corners of the shapes. All of the shapes are well known.

As a help, here are 2 examples.

Example

(i) Can you find these shapes:

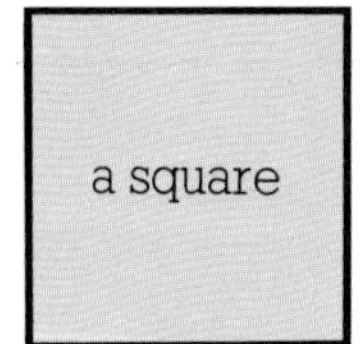

and

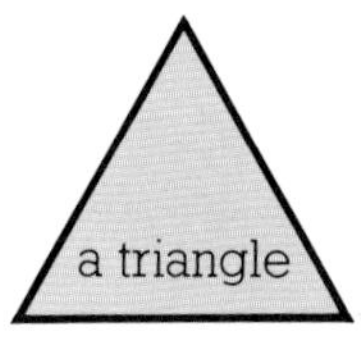

in these crosses?

This should not be too difficult.

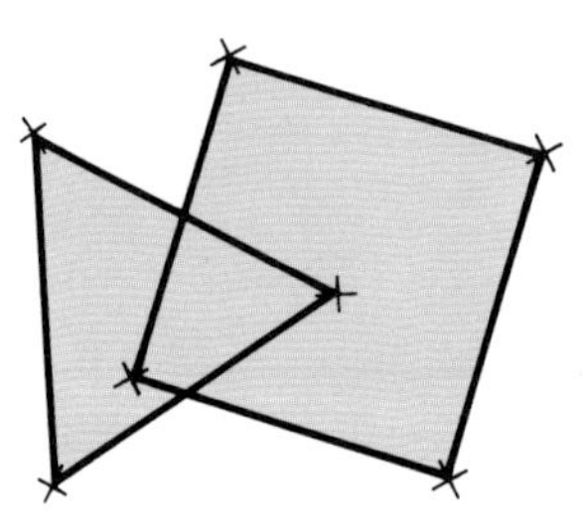

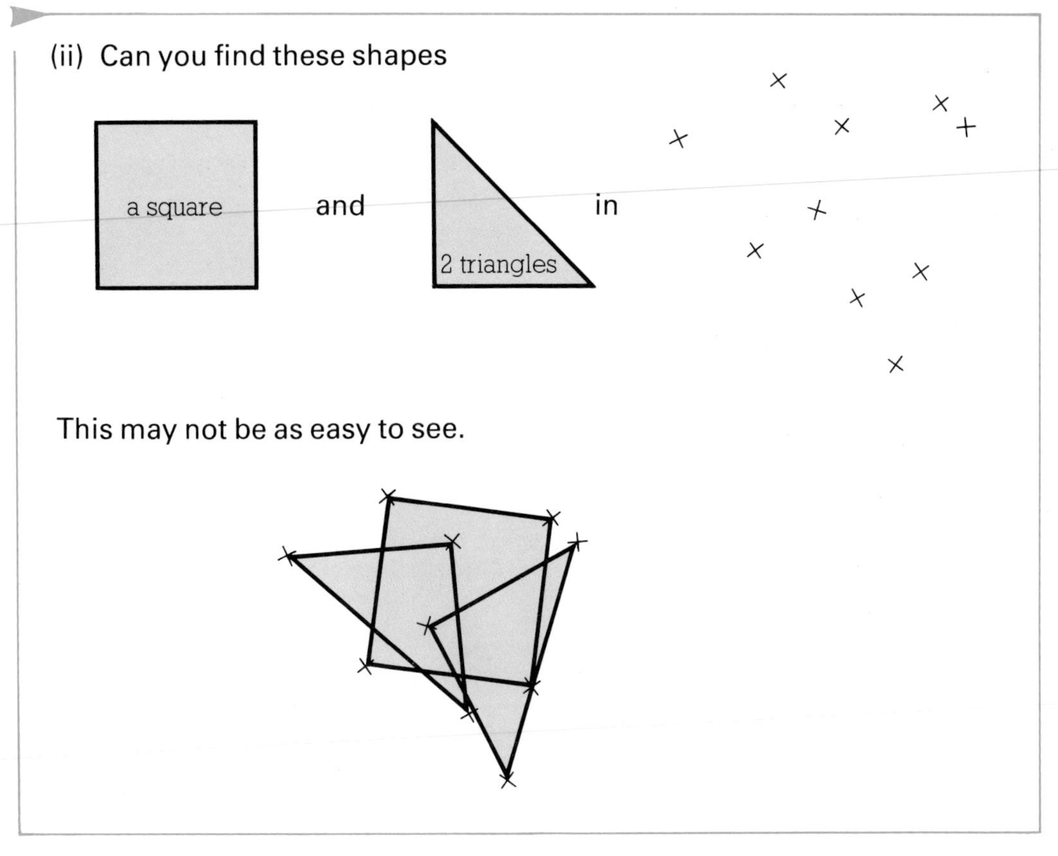

To do the questions use tracing paper to copy the cross pattern onto paper or into your exercise book.

Write down any ideas you might have about systematic ways of finding the hidden shapes.

Questions 1–6

Find the shapes shown in the image crosses.

1

2 squares

1 triangle

2

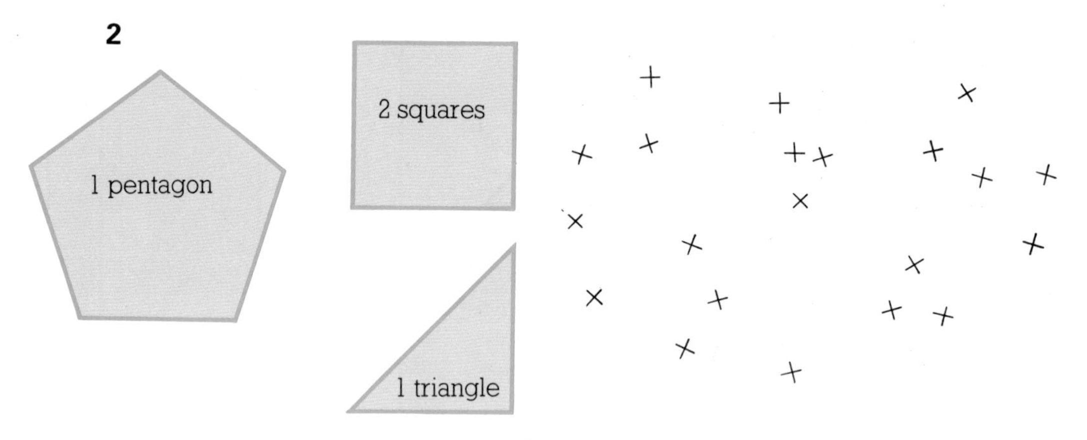

3 All 3 shapes share a common point.

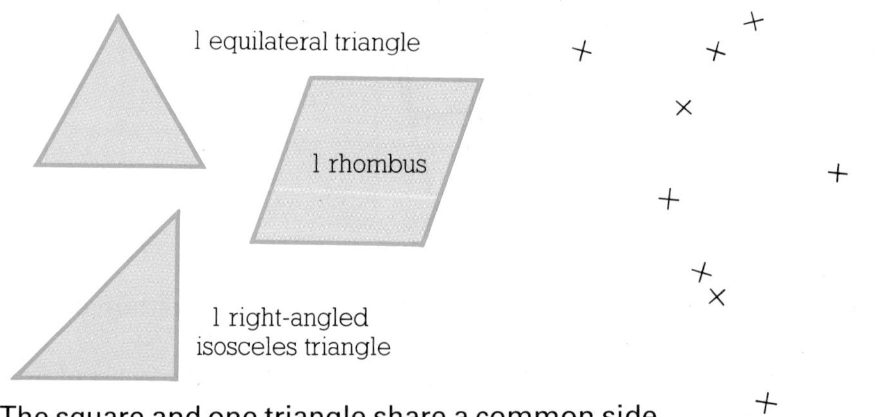

4 The square and one triangle share a common side.

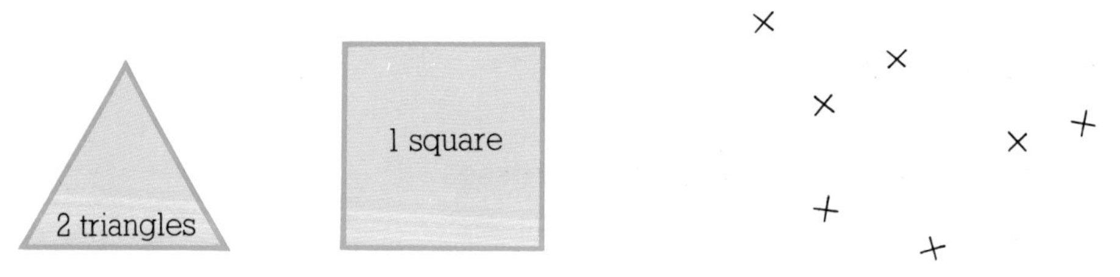

5 Two extra dots have been included which are not on any of the shapes.

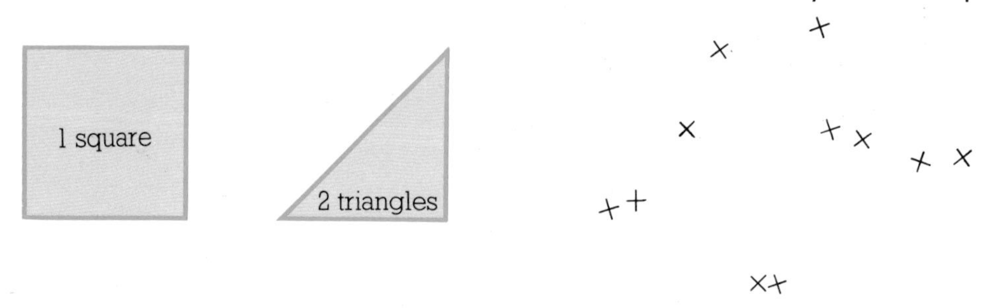

6 Draw in the shapes you can find.

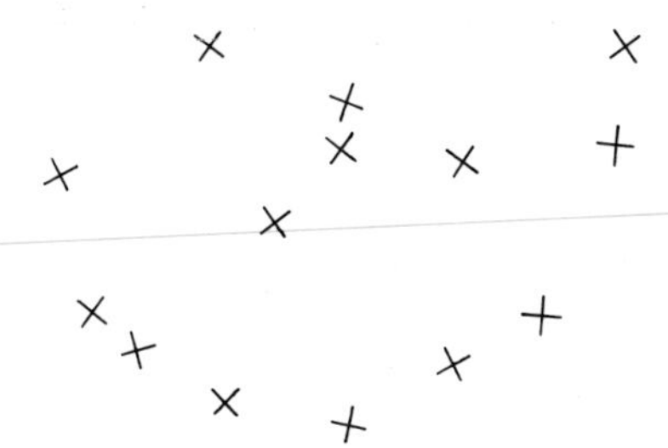

Question 7

Using shapes of your own, make up 5 hidden dot shape problems for your neighbours to solve.

21 Fraction Activities

Example

By adding 3 to the numerator and denominator of $\frac{1}{3}$, the fraction is doubled.

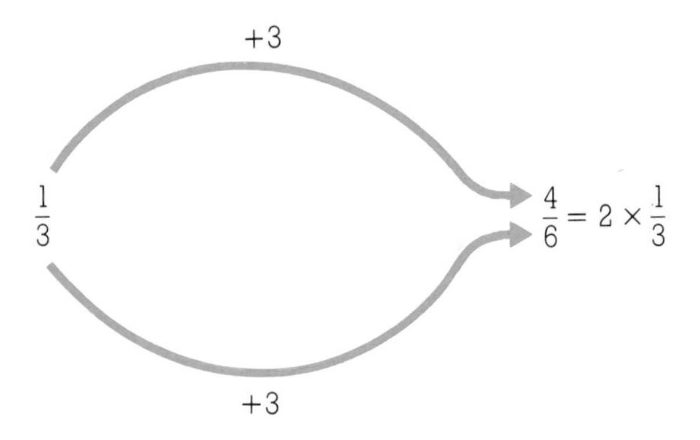

Question 1

What number added to both **numerator** and **denominator** is needed to double these?

1 $\frac{1}{4}$

2 $\frac{2}{5}$

3 $\frac{3}{7}$

4 $\frac{3}{8}$

5 $\frac{4}{9}$

6 $\frac{5}{11}$

7 $\frac{1}{2}$

8 $\frac{4}{7}$

Can you do them all? Try the same thing for trebling.

Question 2

If fractions are plotted like this:

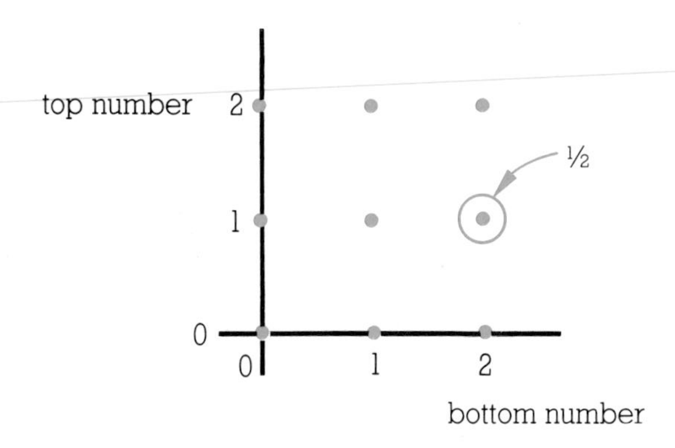

how many *different* **non-equivalent** fractions will there be on a 10 × 10 grid? (A fraction is less than 1.)

How many different fractions will there be on a 20 × 20 grid?

What do you notice about the sets of equivalent fractions?

Question 3

Start with a fraction, e.g. ⅘. Work out:

$$\frac{1 - \text{fraction}}{1 + \text{fraction}}$$

$$\frac{1 - \frac{4}{5}}{1 + \frac{4}{5}} = \frac{\frac{1}{5}}{\frac{9}{5}} = \frac{1}{9}$$

Do the same again. What do you get?

Try it with several fractions.

Question 4

Investigate the length of cycle for recurring decimals such as:

$$\frac{1}{11} = 0.090909 \text{ etc.}$$

Now try the following:

$$\frac{1}{3}, \frac{1}{7}, \frac{1}{11}, \frac{1}{13}, \frac{1}{17}, \frac{1}{19}, \frac{1}{23}$$

Question 5

What fraction is doubled when 2 is added to the numerator and denominator?

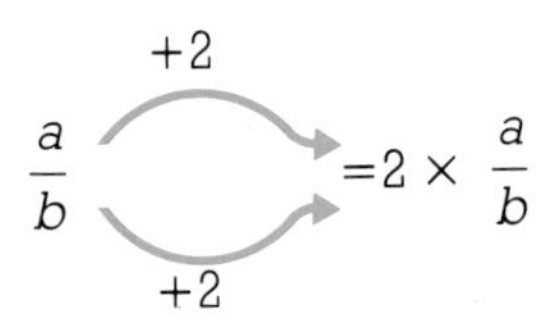

Question 6

If

$$\frac{1}{a}+\frac{1}{b}=\frac{2}{3}$$

What values can a and b have? Give more than one answer.

Question 7

Find x and y such that

$$\frac{1}{x}-\frac{1}{y}=\frac{2}{7}$$

and show that

$$\frac{x}{y}=\frac{1}{7}$$

Question 8

Choose any number, say 7. Subtract the reciprocal from 1, i.e.

$$1-\frac{1}{7}=\frac{6}{7}$$

Now subtract the reciprocal of $\frac{6}{7}$ (which is $\frac{7}{6}$) from 1:

$$1-\frac{7}{6}=-\frac{1}{6}$$

Keep going; what happens?
Do the same using other starting numbers.

Question 9

Choose a number x. Work out

$$\frac{1 + x}{1 - x}$$

Call this new value x. Keep repeating the process. What happens?

In question 10 use your calculator and key in

a single digit *then* ÷ *then* another number *then* =

and you must try to get each of the numbers shown below.

> ### *Example*
>
> To get 0.4, you key in
>
> 4 *then* ÷ *then* 10 *then* = 0.4

Question 10

A dot above a number means it is recurring e.g. $0.\dot{4} = 0.44444$ etc.

0.1	$0.8\dot{3}$	0.4	0.375	0.6
$0.5\dot{5}$	$0.1\dot{1}$	0.8	$0.2\dot{2}$	0.625
0.5	0.25	$0.7\dot{7}$	$0.6\dot{6}$	0.2
0.875	$0.3\dot{3}$	0.75	$0.4\dot{4}$	0.125
0.3	$0.8\dot{8}$	0.7	$0.16\dot{6}$	0.9

Question 11

Gaps in a piece of machinery are checked by using metal strips of standard thickness.

There are 4 strips of thickness ½, ¼, ⅛ and $\frac{1}{16}$ of a centimetre. To check a gap of ⅜, the ¼ and ⅛ centimetre strips are used.

Suppose the ⅛ centimetre strip is lost.

(i) Make a list of all the gaps which could be measured using the ½, ¼, and $\frac{1}{16}$ centimetre strips.

(ii) An extra set of strips is made of thicknesses 0.4, 0.2, and 0.1 centimetres. There are now 6 strips. Place them in order of size.

(iii) How many different thicknesses can be measured using these strips? What is the largest thickness that can be measured?

22 Number Bases 2
Adding and Subtracting

Adding

For this method you need counters and properly spaced circles.

Example

In base 2

$$\begin{array}{r} 101+ \\ \underline{11} \end{array}$$

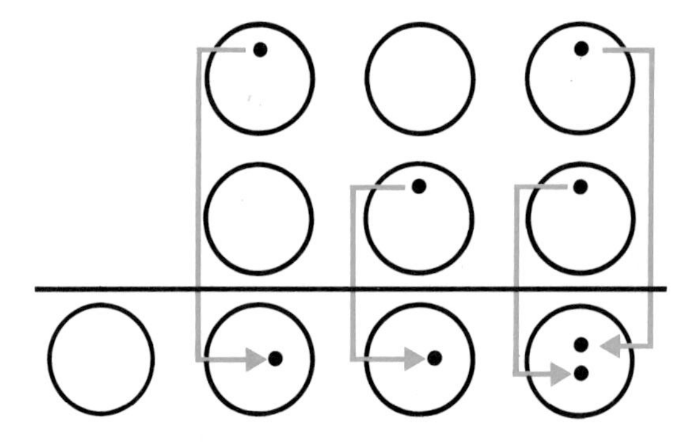

Now, starting from the right-hand units column and working towards the left, group the counters. Remember that the problem is in base 2 so a group of 2 counters will be worth 1 in the next column along.

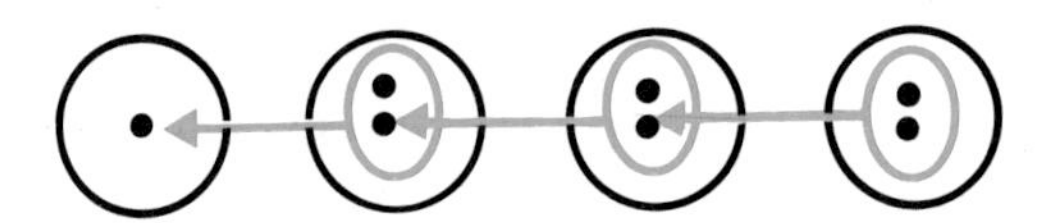

Answer: $1000_{(2)}$

Question 1

Answer the following questions
In base 2:

1 100+ 11	**2** 101+ 1	**3** 110+ 11	**4** 1111+ 111
5 1010+ 1010	**6** 1011+ 1011	**7** 11110+ 1011	**8** 10011+ 11101

In base 3

9 222+ 222	**10** 12+ 12 12	**11** 212+ 101 12	**12** 111+ 202 102

In base 5:

13 41+ 34	**14** 1022+ 1022	**15** 444+ 444	**16** 14312+ 2312

Now try some more of your own.

Subtracting

Use a similar method for subtraction, making sure you have enough counters available in the top circles.

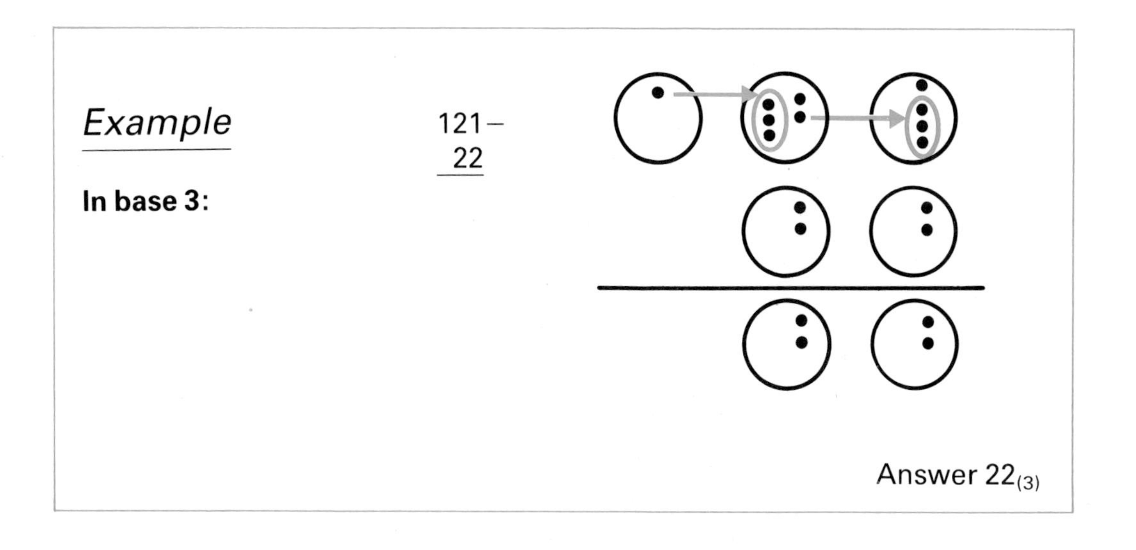

Question 2

Now try the following questions:
In base 2:

1 11− 1	**2** 10− 1	**3** 101− 11
4 1011− 101	**5** 10000− 1101	**6** 11011− 11011

In base 3:

7 112− 21	**8** 102− 20	**9** 1000− 222
10 101011− 1110	**11** 2111− 222	**12** 1011− 201

In base 5:

13 14− 3	**14** 13− 4	**15** 101− 44

In base 10:

16 9716− 888	**17** 10000− 4769	**18** 92157− 2892

Now try some more of your own.

23 Pythagoras 1

The diagrams below all show squares drawn on the sides of triangles. The triangles are named according to the type of angles they have within them:

(i) all acute
(ii) obtuse
(iii) right-angled.

The areas of all the squares are shown.

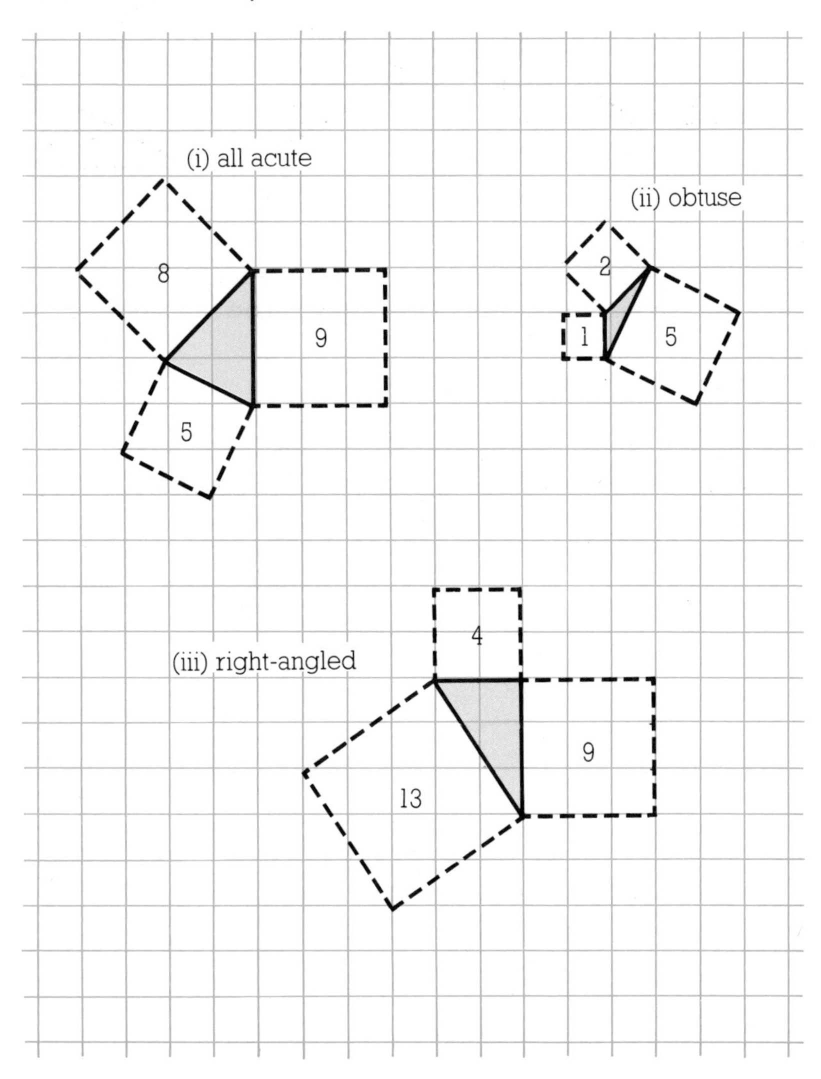

Question 1

The diagrams show some triangles drawn on squared paper.

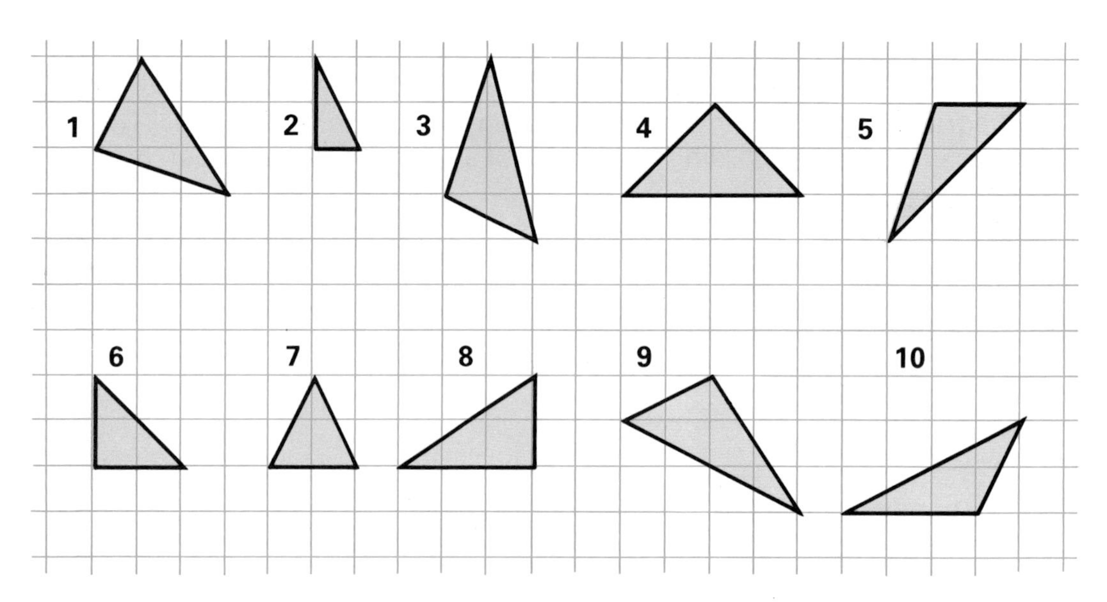

Copy each diagram onto your own squared or dotted paper.
Identify which type each triangle is and draw squares on the sides of each triangle.
Work out the area of each square.

Set out your results in a table like the one below, which has been completed for the three triangles at the beginning of the chapter.

type	*largest area*	*middle area*	*smallest area*
all acute	9	8	5
obtuse	5	2	1
right-angled	13	9	4
1			
2			
⋮			
10			

When you have completed all of these, your table should contain 13 sets of results.

Question 2

What can you say about the angles in the triangle in the cases:

(i) where the largest area is less than the sum of the other two
(ii) where the largest area is greater than the sum of the other two
(iii) where the largest area is equal to the sum of the other two?

Set out all your results and conclusions clearly.

24 Pythagoras 2

All lengths in this chapter are in centimetres and all areas are in square centimetres unless it says otherwise.

Question 1

In each of the diagrams below the triangles are right-angled. Squares have been drawn on each of the sides and the *areas* of these squares are written inside some squares. However, one area is not known and is marked with a '?'. Calculate the value of this unknown area.

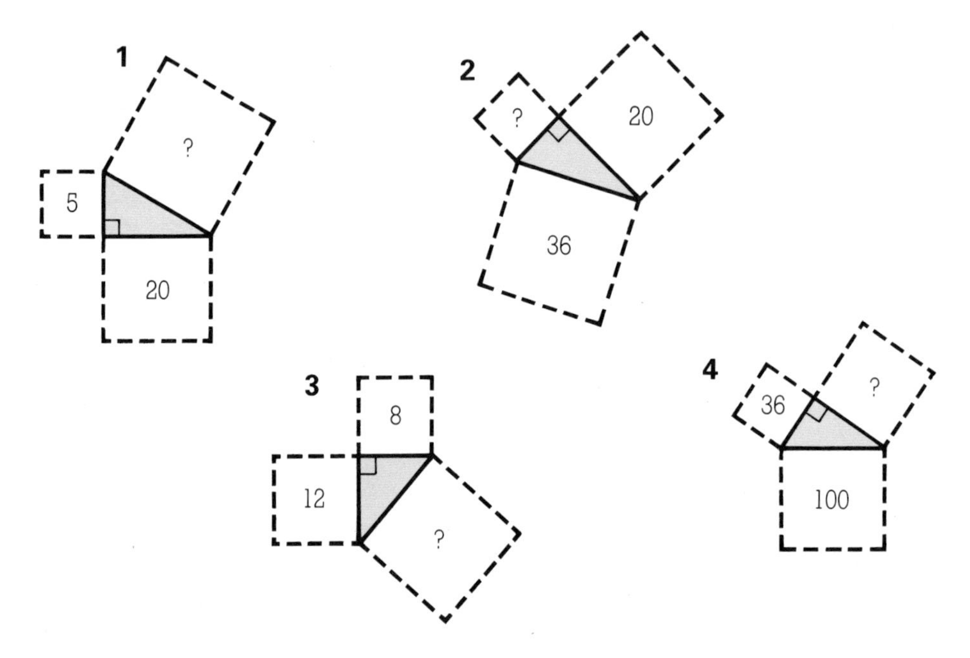

Question 2

You are given two of the lengths of the sides of a right-angled triangle. Calculate the *area* of the square drawn on the third side.

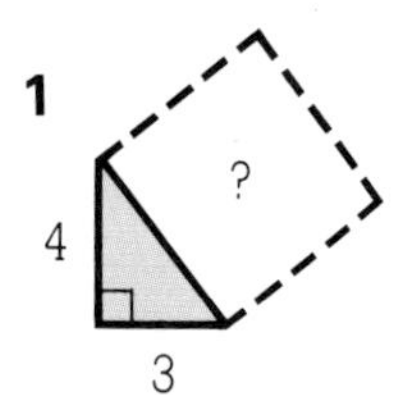

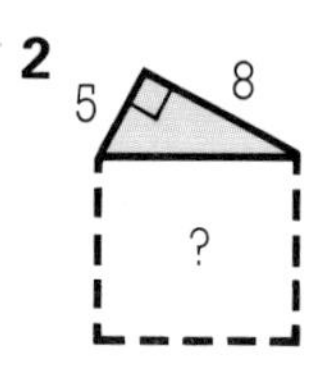

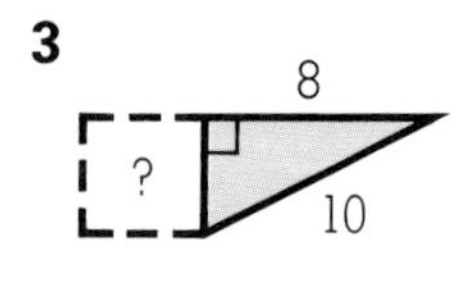

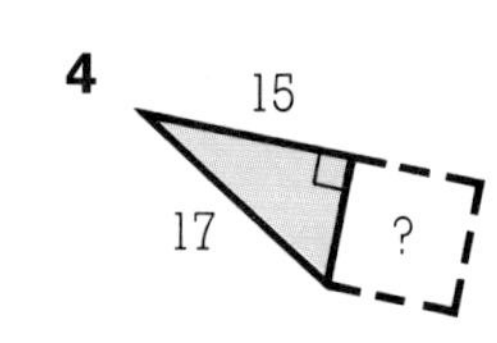

Question 3

The sketches below all show triangles which have two sides of known length and one of unknown length x.
Calculate the value of x in each case.

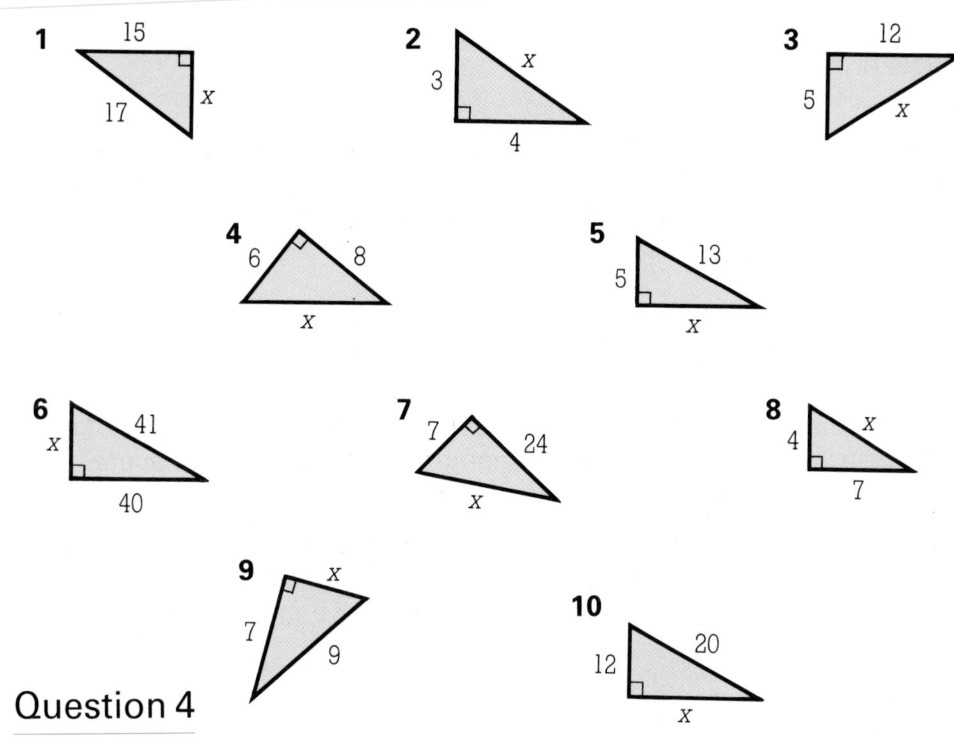

Question 4

Calculate the lengths of the sides marked x and y in each case below.

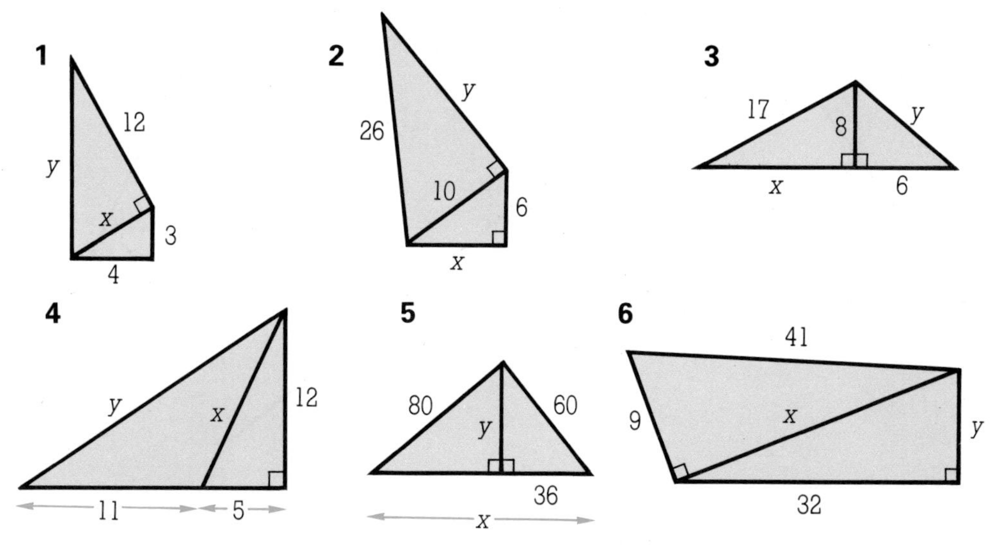

Question 5

A builder uses a single length of rope to mark out a triangular plot of land X Y Z. The length XY = 16m, the length YZ = 30m and the angle at Y is 90°.
Calculate the total length of rope that the builder will need.

Question 6

One end of a ladder of length 12 metres rests against a vertical wall. The other end of the ladder is on horizontal ground and is exactly 7 metres from the base of the wall. Calculate the height of the highest point of the ladder from the horizontal.

Question 7

In each of the following cases, the numbers in brackets are the lengths of the sides of some triangles. Separate these triangles into groups of *all acute, obtuse* or *right-angled*. Show all your working.

1 (7, 8, 11)
2 (8, 6, 10)
3 (5, 12, 13)
4 (8, 15, 19)
5 (7, 8, 10)
6 (7, 2, 8)
7 (6, 6, 10)
8 (17, 8, 15)
9 (9, 41, 40)

Question 8

Calculate the lengths of the lines AB, CD, DE and FG below. Show all your working.

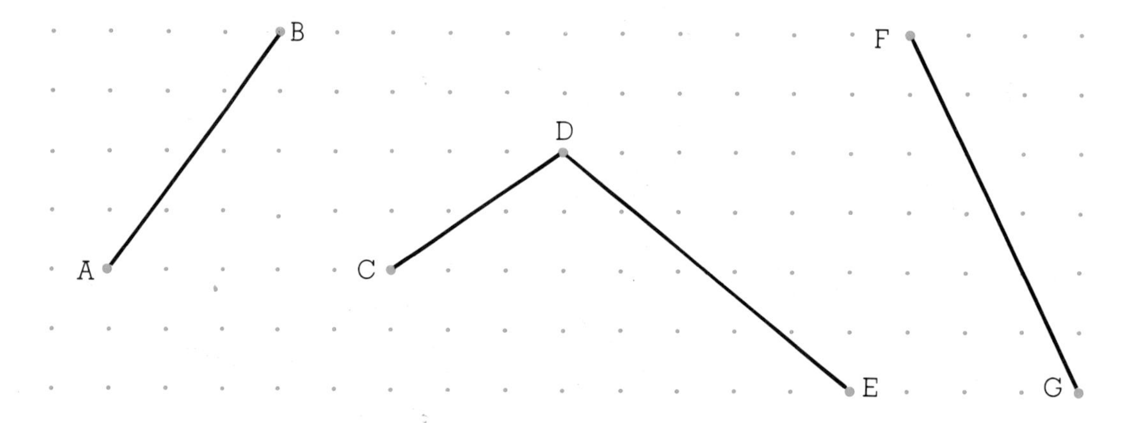

Question 9

On graph paper plot the points:

$$A = (3, 1) \quad \text{and } B = (7, 4)$$

$$P = (-1, 3) \text{ and } Q = (11, 8)$$

Calculate the distances:

1 AB **2** PQ

Question 10

On graph paper plot the points:

$(5, 0)$; $(4, 3)$; $(3, 4)$; $(0, 5)$; $(-3, 4)$;
$(-4, 3)$; $(-5, 0)$; $(-4, -3)$; $(-3, -4)$; $(0, -5)$
$(3, -4)$; $(4, -3)$.

Use Pythagoras' theorem to prove that all the points lie on a circle with centre $(0, 0)$. What is the radius of that circle?

Question 11

The diagonal of a square is exactly 10 centimetres in length. Calculate the lengths of the sides of the square.

Question 12

The dimensions of a cuboid are $6 \times 8 \times 11$ centimetres. Showing all of your working, calculate the length of the longest diagonal of the cuboid. It may be helpful to draw a diagram of the cuboid.

25 Geoboard 3
Shapes

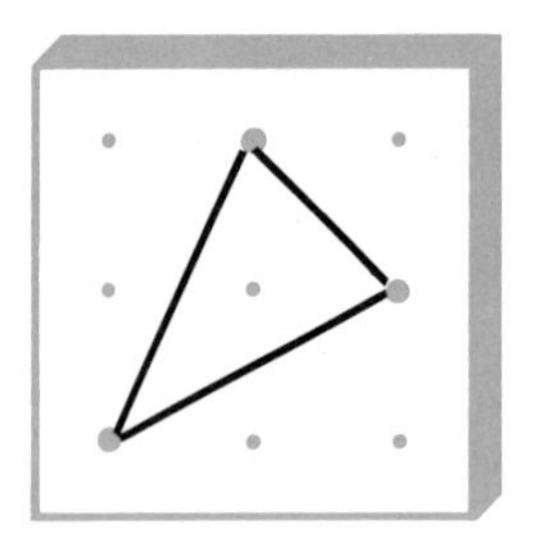

Question 1

On the 9 pin geoboard how many *different* triangles can you make? Draw your triangles on dotted paper. Do any of the triangles have special names? Record your results.

Question 2

On the 9 pin geoboard how many *different* quadrilaterals can you make? Draw all of your quadrilaterals on dotted paper. Do any of the quadrilaterals have special names? Record your results.

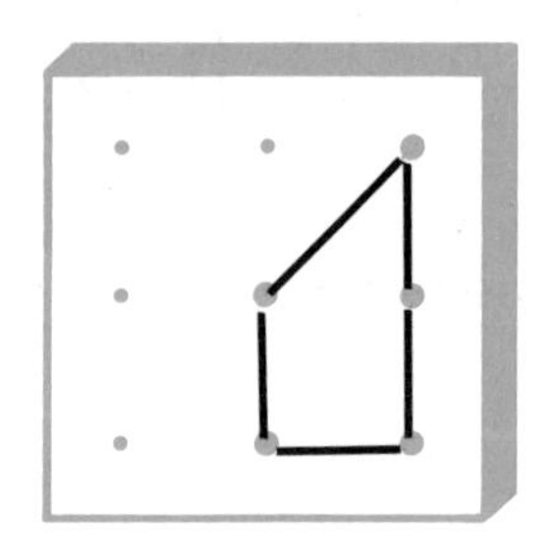

Question 3

Using a 16 pin geoboard, can you make:

1 5 different sized squares
2 a rhombus that is not a square
3 9 different parallelograms (not rectangles)
4 4 different kites
5 4 different rectangles
6 5 (at least) different trapezia
7 5 different arrowheads
8 the polygon with the largest possible number of sides?

In question 4 you are asked to investigate the largest number of sides a polygon can have on an $n \times n$ geoboard (with no cross-over figures and only 1 interior region allowed).

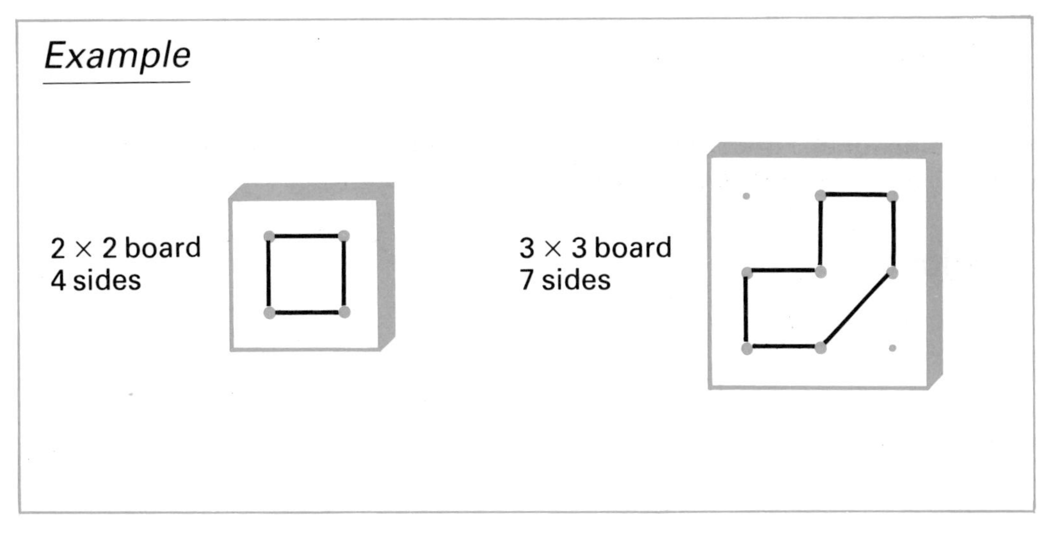

Question 4

Find the polygon with the largest number of sides that can be made on these geoboards:

4×4
5×5
6×6
etc.

Draw your results on dotted paper, showing the largest polygon for each board.

Make a table of your results.

Can you say what is the largest number of sides a polygon can have on an $n \times n$ geoboard?

In question 5 you will investigate the largest **convex** polygon on a geoboard.

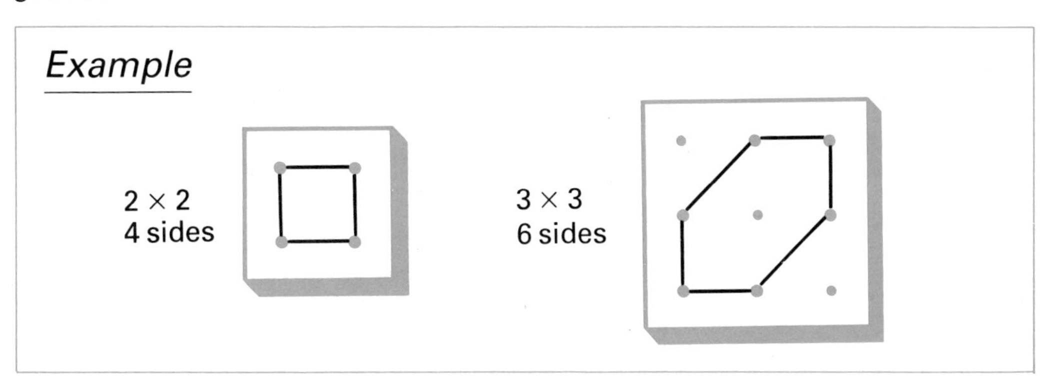

Question 5

Do the same for this question as you did for question 4 but look at **convex** polygons as shown in the example.

Question 6

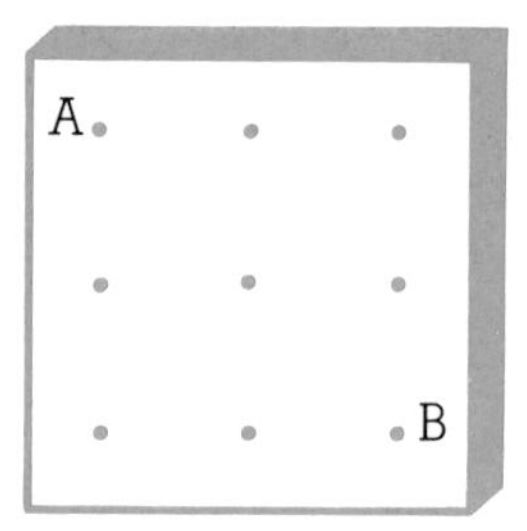

There are many routes from A to B which pass through each pin *exactly* once.

Find:

1 the shortest route
2 the longest route where the stages do *not* cross
3 the longest route where crossing *is* allowed.

26 Cuisenaire Activity 4
Equivalent forms

w : *white*
r : *red*
g : *light green*
p : *pink*
y : *yellow*
d : *dark green*
b : *black*
t : *tan* (brown)
B : *blue*
O : *orange*

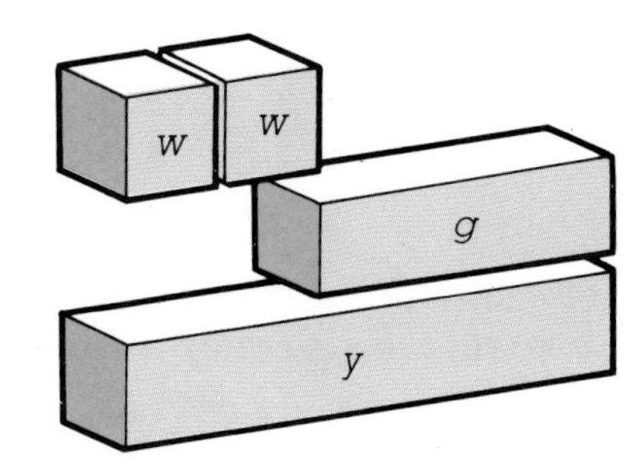

Lay out the *red* and *green* Cuisenaire rods end to end as a rod train:

r	g

Do this again so you have all 4 rods lying end to end as another rod train:

r	g	r	g

This is 2 lots of (*red* + *green*) or $2(r + g)$

We can shuffle the rods around in many ways. For instance:

r	r	g	g

$r + r + g + g$ or $2r + 2g$

g	r	r	g

$g + 2r + g$

However since these rod trains all use the same 4 rods, we can say that they are equivalent or **equivalent forms**.

So we can write:

$$\begin{aligned} 2(r + g) &= r + r + g + g \\ &= 2r + 2g \\ &= g + 2r + g \end{aligned}$$

Question 1

Write down as many other equivalent forms to $2(r + g)$ as you can.

Question 2

Set up each of the following with the rods. Then, in each case set up and write down as many equivalent forms as you can for each one.

1 $2(g + p)$
2 $3(g + y)$
3 $3(2w + g)$
4 $2(3r + 2p)$
5 $3(g + 2p + 3r)$

We can do something similar even when we have subtraction signs. For instance, *yellow* minus *red* is set up as follows:

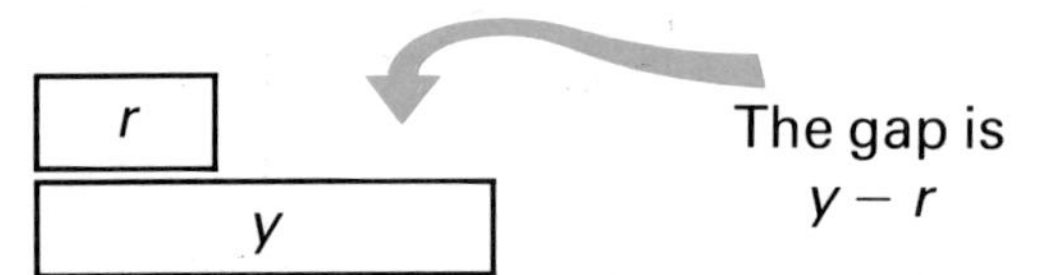

Doing this twice we can have:

r $y - r$ r $y - r$
y   y

The total gap is $(y - r) + (y - r)$ or $2\,(y - r)$.

Moving a *red* rod across we can have:

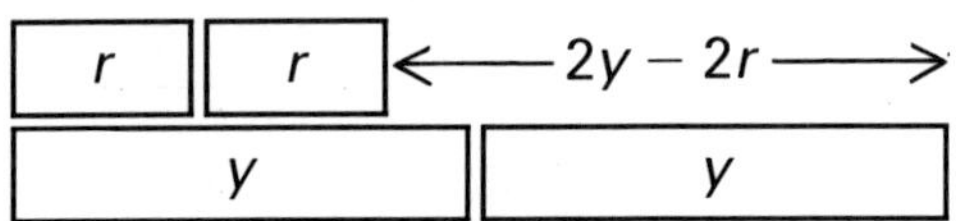

or that gap could be

r   r $y - 2r$ + y
y   y

So, since all the gaps are of the same length, we can say that

$$(y - r) + (y - r)$$
$$2\,(y - r)$$
$$2y - 2r$$
$$y - 2r + y$$

are all equivalent forms.

Question 3

Can you find any more equivalent forms to:

$$2\,(y - r)$$

Write them down if you can.

Question 4

Set up each of the following with the rods. Then in each case set up and write down as many equivalent forms as you can for each one.

1 $2\,(b - p)$
2 $3\,(y - r)$
3 $2\,(2g - r)$
4 $3\,(2y - g)$
5 $3\,(4y - 3g)$

Multiplying out and factorising

You have seen that

$$2\,(r + g) = 2r + 2g$$

If we go from

$$2\,(r + g) \quad \text{to} \quad 2r + 2g$$

it is called **multiplying out**.

If we go from

$$2r + 2g \quad \text{to} \quad 2\,(r + g)$$

it is called **factorising**. These are special equivalent forms.

You may or may not feel the need to use rods for the next two questions. The choice is yours.

Question 5

Multiply out:

1 $3\,(y + b)$
2 $2\,(3p + w)$
3 $4\,(2y + B)$
4 $3\,(g + w)$
5 $3\,(4w - g)$
6 $5\,(3p - y)$
7 $4\,(3b + 2g)$
8 $3\,(2y + r - g)$
9 $5\,(3t - 2b)$
10 $4\,(3p + 2w - 3g)$

Question 6

Factorise:

1 $2g + 2w$
2 $3g - 3r$
3 $3b - 6w$
4 $4g + 2w$
5 $3t + 9r$
6 $4y + 6p$
7 $5y - 5w$
8 $6g + 9w$
9 $2p + 4g + 6r$
10 $3y - 6g + 3p$

Equivalent forms without rods

The same principles of equivalent forms, multiplying out and factorising, hold true for general letters and numbers which may or may not represent the rods.

For instance, all of these are equivalent forms:

$$\begin{aligned} 2(x + y) &= 2x + 2y \\ &= x + 2y + x \\ &= x + x + 2y \end{aligned}$$

and so on.

Question 7

Write down as many *equivalent forms* as you can for each of these:

1 $2(x + y)$
2 $3(x + y)$
3 $2(3x + y)$
4 $3(2x - y)$
5 $5(2x + 3y)$
6 $x + 2y + 3x + 5y$
7 $2x + 3y - x - y$
8 $3y + 7x - y - 3x$
9 $x + y + 4x - 2y + 2y + 3y$
10 $3x - y + 2x + 6y$

Before doing the remaining question, it may be helpful to remember that:

we write	$3 \times x$	as $3x$
that	$x \times x$	is x^2
that	-2×-3	is $+ 6$ (or 6)
that	$4 \times 3x$	is $12x$

and similar.

Question 8

Multiply out:

1 $3(x+y)$
2 $2(2x+6y)$
3 $3(2x-y)$
4 $5(x+6y)$
5 $4(x+3)$
6 $7(2-3x)$
7 $4(2x-3)$
8 $-3(x+y)$
9 $-2(-3x-2)$
10 $-4(-3-2y)$
11 $-2(3x-2y)$
12 $x(x+2)$
13 $y(y+3)$
14 $x(3x+4)$
15 $x(x-4)$
16 $y(3-y)$
17 $(2-3x)$
18 $x(1-3x)$
19 $y(1-y)$
20 $2x(3-x)$

Question 9

Factorise:

1 $2x+4y$
2 $3x-2y$
3 $2x-6$
4 $4y+8$
5 $9-3x$
6 $5x+15$
7 $3x+21$
8 $4-2y$
9 $6x+9$
10 $3y+12$
11 $x^2-6x)$
12 y^2+4y
13 $5y+y^2$
14 $3x-6x^2$
15 $4x-x^2$
16 $5y+y^2$
17 $5x^2-10x$
18 $4x^2+12x$
19 $6y^2-2y$
20 $4y-6y^2$

27 Number Bases 3
Multiplying and dividing

Multiplying

First set the number to be multiplied up in the *top* set of circles and begin on the *right*. Then convert the number you are multiplying by to its base 10 equivalent.

Example

Work out in base 2 $\begin{array}{r} 101\times \\ \underline{11} \end{array}$

101 is arranged in the top set:

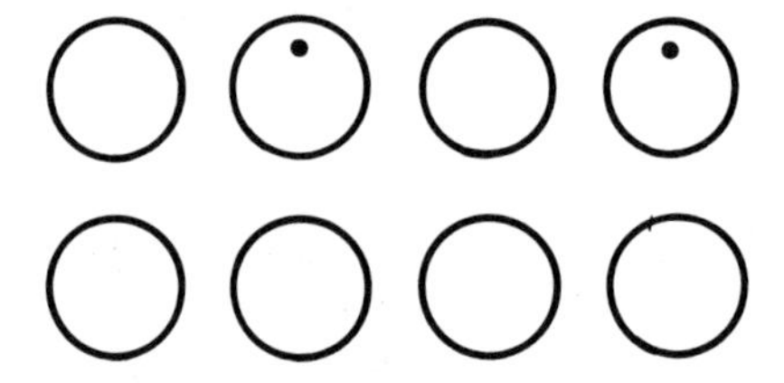

We know $11_{(2)} = 3_{(10)}$ so, starting at the right-hand end we have:

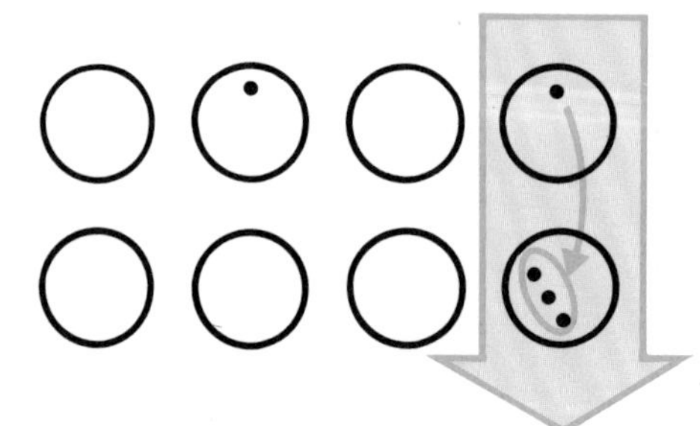

Now we convert into base 2.

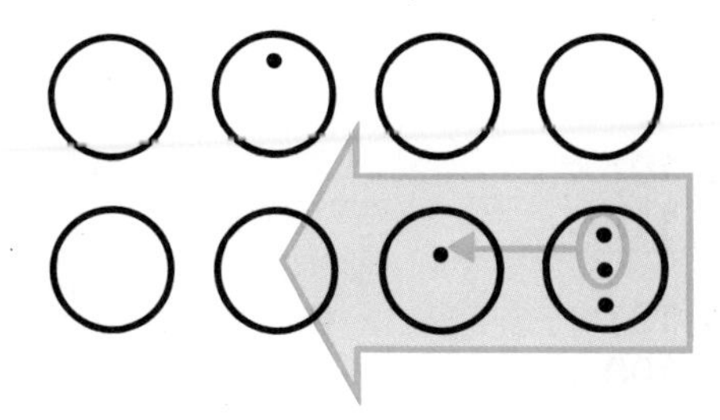

There is nothing to multiply in the second column, so move to the third.

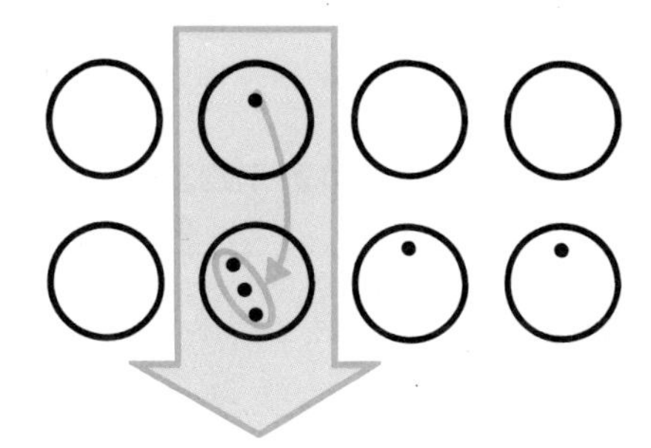

Again this needs converting.

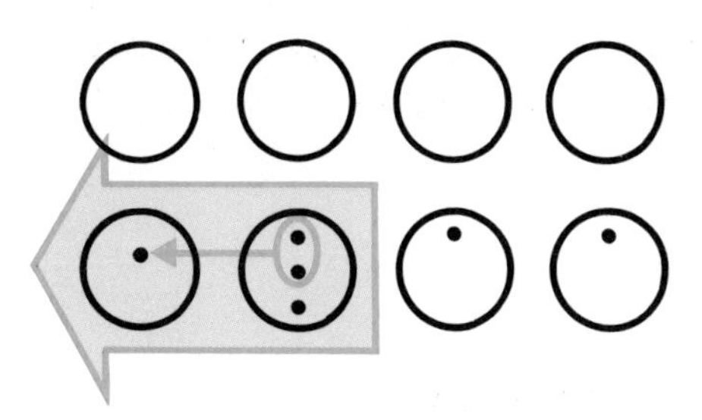

Answer: $1111_{(2)}$

Check: by converting back to base 10

$101_{(2)} = 5_{(10)}$
$11_{(2)} = 3_{(10)}$
$5 \times 3 = 15_{(10)} = 1111_{(2)}$

Question 1

Now try the following:
In base 2:

1 $\begin{array}{r} 1011\times \\ \underline{11} \end{array}$ **2** $\begin{array}{r} 101\times \\ \underline{110} \end{array}$ **3** $\begin{array}{r} 110\times \\ \underline{100} \end{array}$

4 $\begin{array}{r} 111\times \\ \underline{10} \end{array}$ **5** $\begin{array}{r} 101\times \\ \underline{100} \end{array}$ **6** $\begin{array}{r} 11011\times \\ \underline{111} \end{array}$

In base 3:

7 $\begin{array}{r} 121\times \\ \underline{22} \end{array}$ **8** $\begin{array}{r} 222\times \\ \underline{12} \end{array}$ **9** $\begin{array}{r} 121\times \\ \underline{10} \end{array}$

In base 5:

10 $\begin{array}{r} 101\times \\ \underline{11} \end{array}$ **11** $\begin{array}{r} 121\times \\ \underline{10} \end{array}$ **12** $\begin{array}{r} 1022\times \\ \underline{4} \end{array}$

Now try some more of your own.

Dividing

This time put the counters in the *bottom* row and start on the *left*. Then convert the number you are dividing by into its base 10 equivalent.

Example

Work out in base 3, 1001 ÷ 21.
1001 is arranged in the bottom row.

We know $21_{(3)} = 7_{(10)}$. Division by 7 is not immediately possible so move counters to the right until enough have been converted to divide by 7.

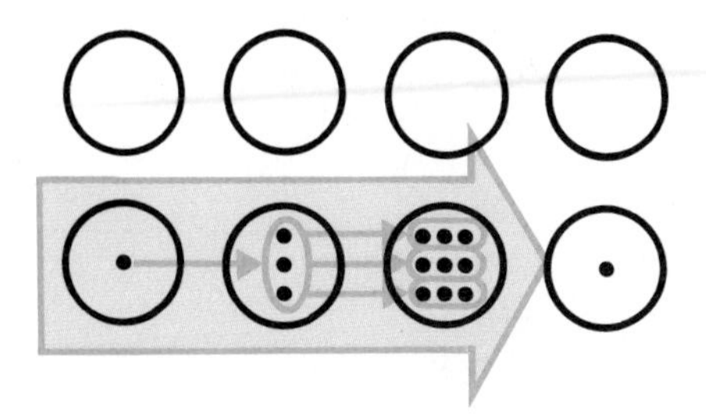

Now divide by 7, by 'taking-out' a group of 7.

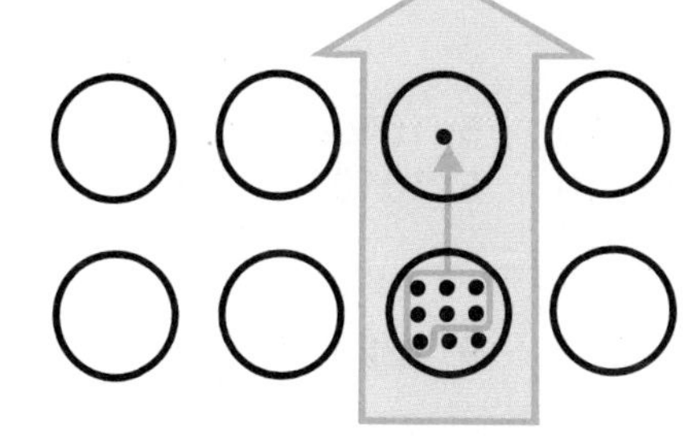

Then *convert* the remaining counters

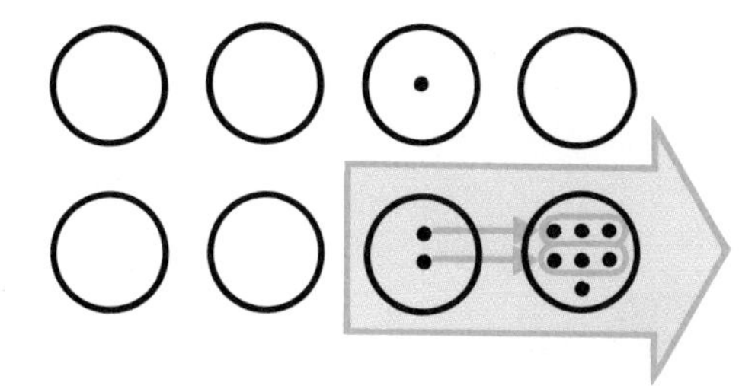

and then divide out

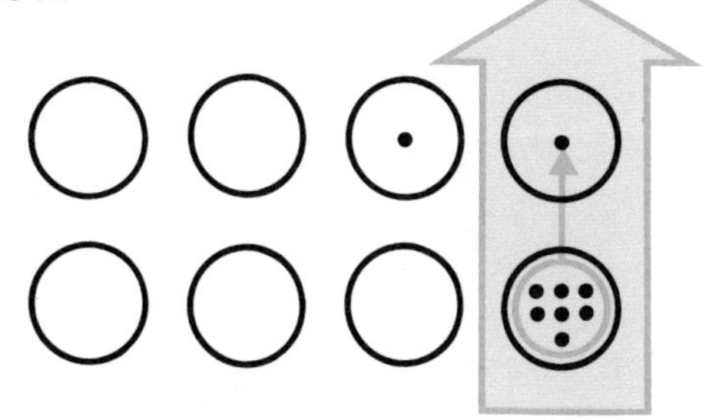

Answer: $11_{(3)}$

Check: by converting back to base 10

$$1001_{(3)} = 28_{(10)}$$
$$21_{(3)} = 7_{(10)}$$
$$28 \div 7 = 4_{(10)}$$
$$= 11_{(3)}$$

Question 2

Now try the following:

In base 2:

1 $1111 \div 11$ **3** $1010 \div 10$ **5** $10100 \div 100$
2 $1010 \div 101$ **4** $1110 \div 10$ **6** $10101 \div 111$

In base 3:

7 $120 \div 12$ **8** $120 \div 10$ **9** $222 \div 111$

In base 5:

10 $41 \div 12$ **11** $30 \div 10$ **12** $121 \div 11$

Now try some more of your own.

28 Area Investigations
Change the shape

Equivalent area shapes can be made by adding areas and then subtracting the same areas from elsewhere — in this case using just squares and triangles.

Example

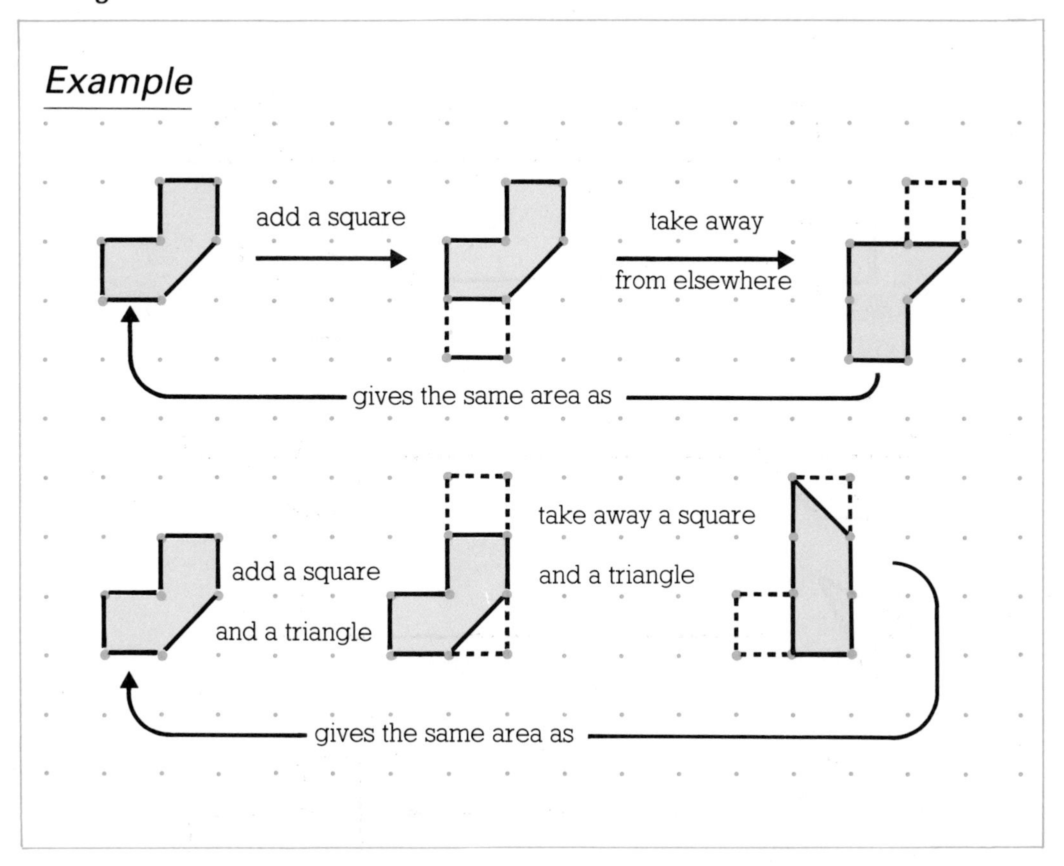

Question 1

Use this method to make the first shape into the second. Illustrate your answers as shown in the examples.

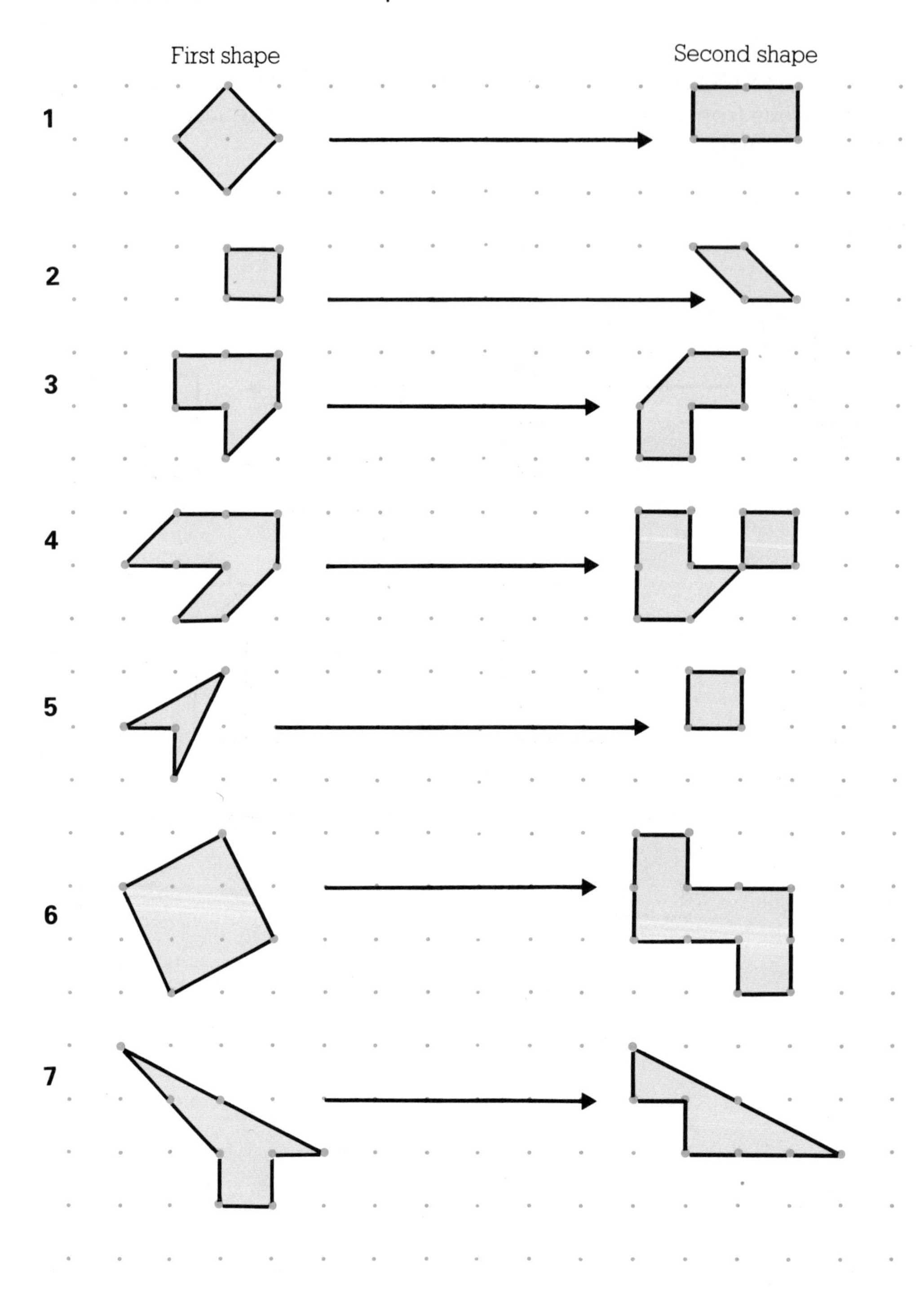

Question 2

Using thick paper or card, make some interesting and unusual shapes by cutting an area of, say, 24 square units (6×4 rectangle). However you cut up the piece, the area always remains the same.

Example

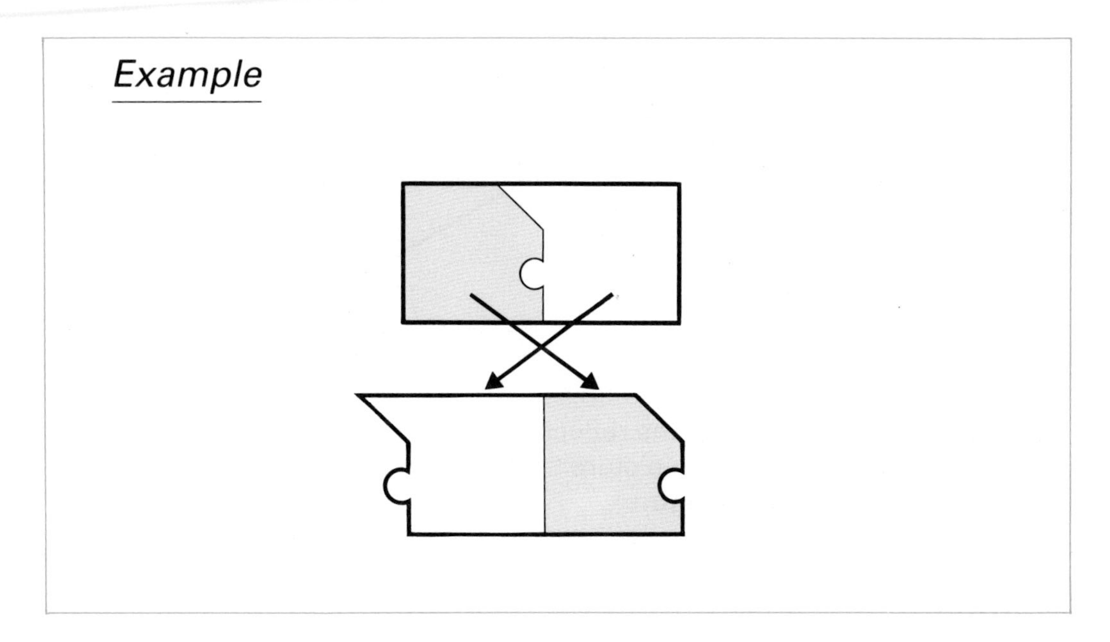

Question 3

Isoperimetric shapes are ones which have the same perimeter.

The perimeter of this shape is 12 units. How many different isoperimetric shapes can you make with a perimeter of 12 units. Write down the area of each shape.

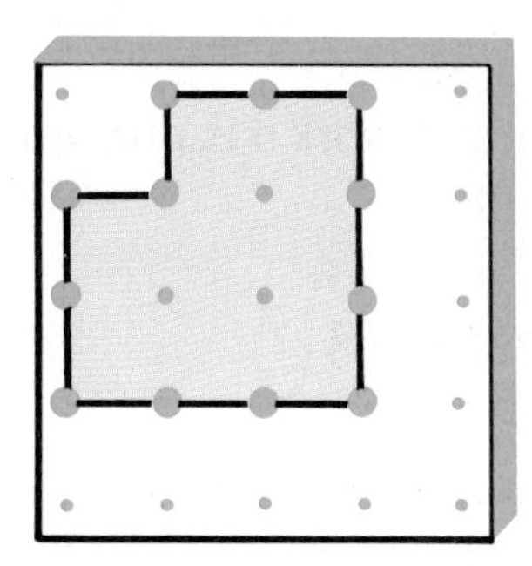

29 Mirror Figures

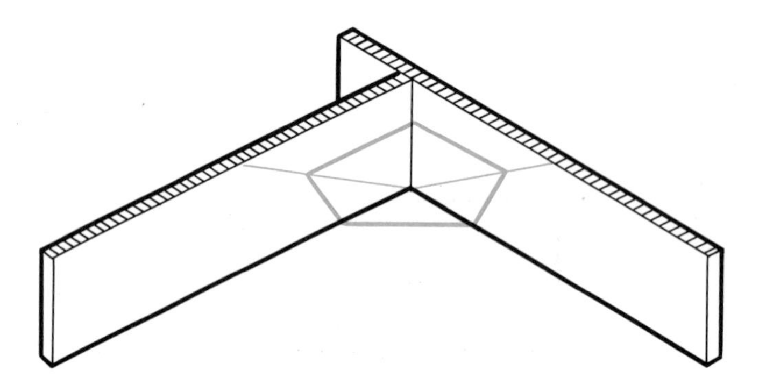

The picture shows a sketch of the reflection of a single straight line in a pair of mirrors held at an angle. In the picture the reflections of the line seem to make a **pentagon**, a 5-sided figure:

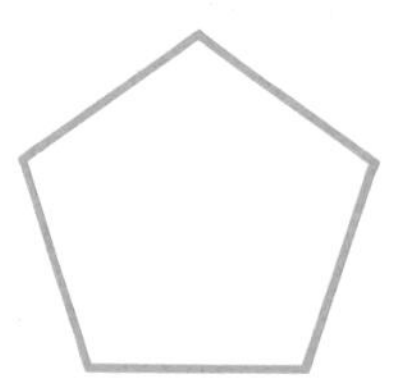

By changing the angle between the mirrors and sliding them backwards and forwards you can change both the shape and size of the figure seen in the reflections.

For this work it is probably best to work in pairs. Share your ideas with your neighbours.

Question 1

Get two mirrors. Draw a single straight line. Spend a little time rotating the mirrors, sliding them backwards and forwards and generally moving them as you like so as to see just what mirror figures you can make.

Try to change the speed at which you rotate the mirrors. You could also tilt the mirrors. Make sketches of the reflected mirror figures. Write some notes on any observations you make.

Question 2

We shall now go back to starting with a single straight line. With this starter, try to make a mirror figure something like each of these:

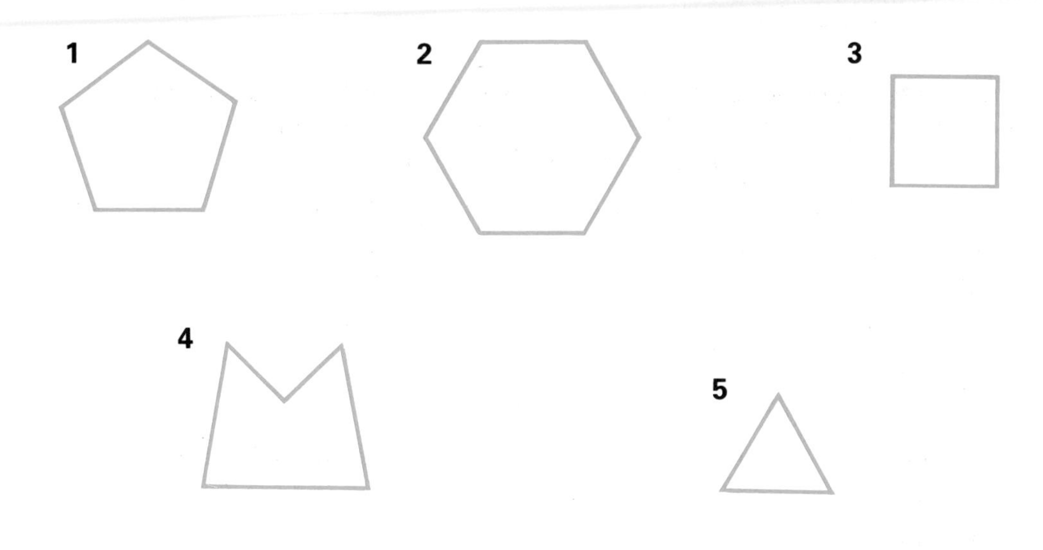

Question 3

It is possible to make a mirror figure which looks like the net of a pyramid. The net is the shape which results when the pyramid is unfolded.

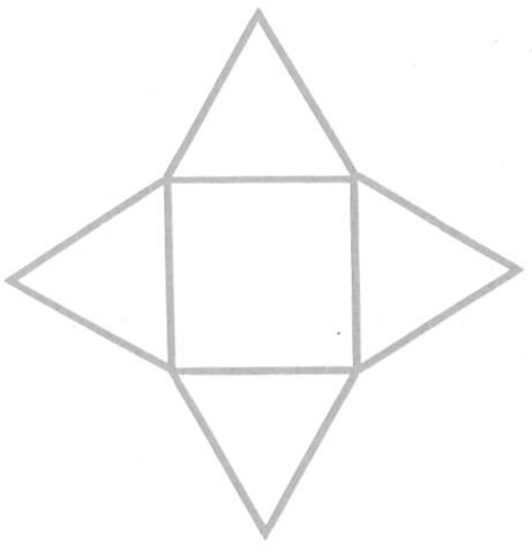

You have to decide what starting shape you will need to have to be able to obtain this mirror figure. There may be more than one starting figure.

Try to make this figure. When you have done so, sketch your starting figure or figures.

Question 4

Experiment with a few starting figures. Then make up a problem similar to question 3 for your neighbours to try to solve. Swap a few problems with these neighbours.

Question 5

A **regular** figure is one in which all the sides are equal in length.

With 2 mirrors and a single straight line as a starting point, try to make *regular* 3, 4, 5, 6, 7 and 8 sided figures. (You will need to do some guessing or estimating of the lengths.)

There is a **general relationship** between the number of sides of the regular figure and the angle between the mirrors. Your task is to work out what this relationship is.

If you need equipment such as protractors then ask. Discuss your ideas with your teacher and neighbours. Write down any observations, results or conclusions.

Question 6

Calculate the angle between the mirrors if you make a regular figure of:

1 5 sides
2 6 sides
3 10 sides
4 100 sides

Question 7

Calculate the number of sides of the regular figure produced when the angle between the mirrors is:

1 120°
2 45°
3 40°
4 18°
5 1°

Question 8

Could you start with a straight line and using two mirrors make a mirror figure which is a circle? Discuss this with neighbours. Write up your conclusion.

30 Manhattan Circle

The streets in Manhattan form a square grid. The buildings in each section are called **blocks**.

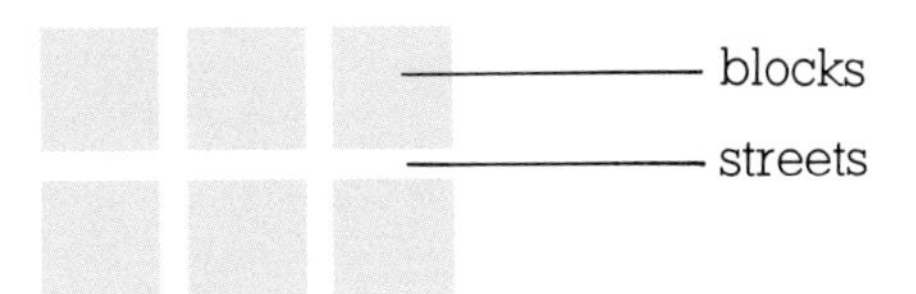

Here is an example of a Manhattan distance:

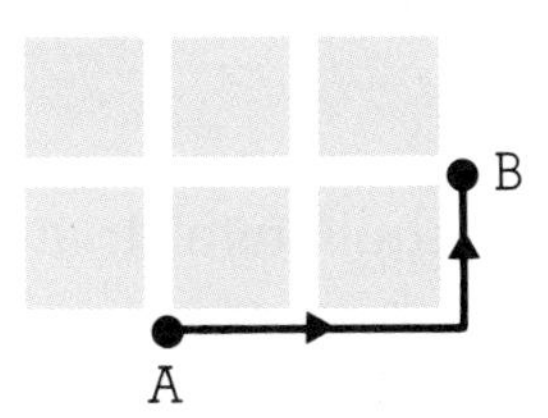

B is 3 blocks away from A.

In our ordinary 2-dimensional geometry, a **circle** is the set of all points which are a constant distance from a fixed point.

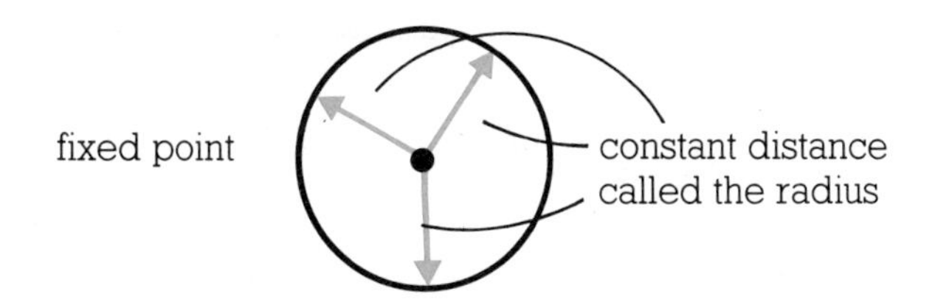

Your task is to see what happens when we transfer this definition to Manhattan geometry. **A Manhattan circle** is the set of all points which are a constant Manhattan distance from a fixed point.

Question 1

On squared paper, mark a fixed point and label it A. Make sure that A is at a crossing of two lines. Mark all the points which are exactly 5 blocks from A. This is a Manhattan circle of radius 5 blocks.

Question 2

Construct a Manhattan circle of:

1 radius 7 blocks
2 radius 8 blocks

Question 3

What do you notice about Manhattan circles? Write some notes on your observations.

Tritown

The streets of Tritown form an equilateral, triangular grid.

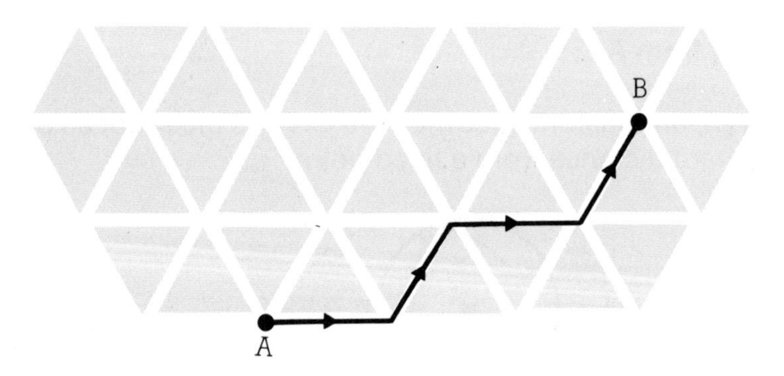

B is 4 triblocks from A.

Question 4

Construct a Tritown circle of radius 5 triblocks.

Question 5

Construct two more Tritown circles of your own choice.

Question 6

What observations can you make about Tritown circles? Write these observations down.

Hextown

In Hextown, the streets form a hexagonal grid.

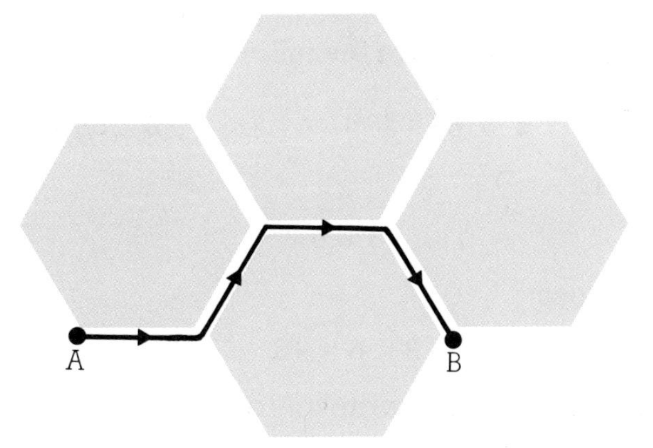

B is 4 hexblocks from A.

Question 7

Investigate some Hextown circles and write down any observations you make about them.

Questrion 8

If you have other lattice paper, use it to examine other circles.

31 Number Theory 2
Last Digit

The **last digit** of 27 is 7.
The last digit of 43218 is 8.
The number 36 is a **perfect square** because $36 = 6 \times 6$.
The number 8 is a **perfect cube** because $8 = 2 \times 2 \times 2$.
In this chapter we shall look at the relationships between last digits and some number sequences.

Question 1

Write down the last digits for each of the following:

1 46
2 124
3 3217
4 4163
5 517
6 268

For the multiplication

$$3 \times 6$$

the last digit is 8, because

$$3 \times 6 = 18$$

and the last digit of 18 is 8. We will write this as

$$\text{LD}\,(3 \times 6) = 8$$

Question 2

Work out each of the following:

1 LD (4×6)
2 LD (3×9)
3 LD (2×7)
4 LD (5×6)
5 LD $(5 \times 7 \times 2)$
6 LD (18×3)
7 LD (4^2)
8 LD (3^4)
9 LD (27×38)
10 LD (426×27)
11 LD $(14 + 17)$
12 LD $(23 + 146)$
13 LD $(3 \times (4+2))$
14 LD $(7 \times (13+8))$
15 LD $(94 \div 2)$
16 LD (4237×1462)
17 LD $(483 - 17)$
18 LD (68^2)

Question 3

By choosing a few values for numbers n and m, try to decide which, if any, of the following are *always* true.

1. $\text{LD}(n+m) = \text{LD}(n) + \text{LD}(m)$
2. $\text{LD}(n^2) = 2 \times \text{LD}(n)$
3. $\text{LD}(n \times m) = \text{LD}(\text{LD}(n) \times \text{LD}(m))$
4. $\text{LD}(n^m) = m \times \text{LD}(n)$
5. $\text{LD}(n \times m) = \text{LD}(n) \times \text{LD}(m)$
6. $\text{LD}(n+m) = \text{LD}(\text{LD}(n) + \text{LD}(m))$

Some, but not all of these results may be of help with the next question.

Question 4

You could work out $\text{LD}(342^3)$ by doing

$$342 \times 342 \times 342$$
$$= 40001688$$

So $\text{LD}(342^3) = 8$

This can be long winded so can you do this in a much simpler way?

Try to do each of these in a simple way. A calculator could be helpful in some cases and as a check.

1. $\text{LD}(47^4)$
2. $\text{LD}(6839 \times 42174)$
3. $\text{LD}(358^5)$
4. $\text{LD}(6473^6)$
5. $\text{LD}(492 \times 3768 \times 487^3)$

Question 5

Construct a table for the last digits of the squares, cubes, fourth and fifth powers of the numbers from 0 to 10.

		LD		
n	n^2	n^3	n^4	n^5
0				
1				
2				
3				
etc.				

Check your results with those obtained by your friends. If you are in any doubt, ask your teacher for some help.

Question 6

There are many patterns which occur in the last digit table you have produced. Explore the table and look for these patterns.

To help you find some patterns, here are a few questions for you to answer.

(i) What happens for numbers greater than 10?
(ii) What can you say about the last digits of:
  (a) perfect square numbers
  (b) perfect cube numbers
  (c) fourth powers
  (d) fifth powers?
(iii) For what values of x would

$$\text{LD}(n^x) = \text{LD}(n)$$

no matter what value n takes?
(iv) For what values of n does

$$\text{LD}(n^x) = n$$

no matter what value x takes?

Feel free to play around with the numbers in the table and offer your own observations.

Question 7

Without doing any long calculation, work out

1 LD (4387^{12})
2 LD (36^{101})
3 LD (73^{62})
4 LD $(57^7 \times 13^8)$

Question 8

If n can be any positive integer, show that

$$10n + 7$$

(i) cannot be a perfect square
(ii) cannot be a perfect fourth power.

Also find *at least* 2 values for n for which

$$10n + 7$$

can be a perfect cube.

32 Special Numbers

If you have worked on investigations such as Handshakes and Line Segments, you will have seen the triangular numbers

$$1, 3, 6, 10, 15, 21, \ldots.$$

In this Chapter we take a much more detailed look at these numbers and some related number patterns.

Triangular Numbers

T_1	T_2	T_3	T_4
$= 1$	$= 1 + 2$	$= 1 + 2 + 3$	$= 1 + 2 + 3 + 4$
$= 1$	$= 3$	$= 6$	$= 10$

Tetrahedral Numbers

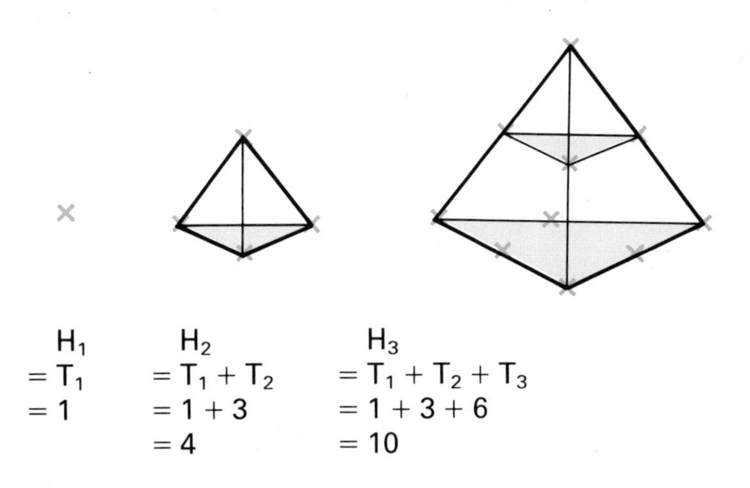

H_1	H_2	H_3
$= T_1$	$= T_1 + T_2$	$= T_1 + T_2 + T_3$
$= 1$	$= 1 + 3$	$= 1 + 3 + 6$
	$= 4$	$= 10$

Square Numbers

S_1	S_2	S_3	S_4
$= 1^2$	$= 2^2$	$= 3^2$	$= 4^2$
$= 1$	$= 4$	$= 9$	$= 16$

Pyramid Numbers

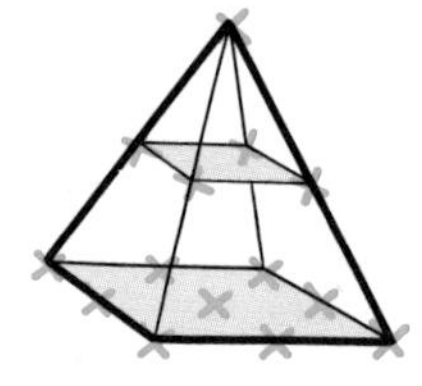

P_1	P_2	P_3	P_4
$= S_1$	$= S_1 + S_2$	$= S_1 + S_2 + S_3$	$= S_1 + S_2 + S_3 + S_4$
$= 1$	$= 1 + 4$	$= 1 + 4 + 9$	$= 1 + 4 + 9 + 16$
	$= 5$	$= 14$	$= 30$

Theorem

$T_3 + T_4 =$

Move the T_3 dots to the top right-hand corner above T_4.

We now have:

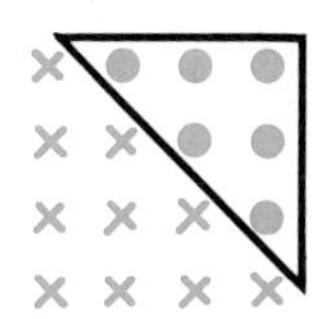

So $T_3 + T_4 = S_4$.

The geometry of the **moving points** is the same in all cases and so the result generalises to:

$$T_n + T_{n+1} = S_{n+1}$$
$$= (n + 1)^2$$

As you do the questions in this chapter, remember that good diagrams will help make your explanations clear.

Question 1

The triangular numbers T_1, T_2, T_3, etc. are calculated as follows:

$$T_1 = 1 \quad T_2 = 1 + 2 = 3 \quad T_3 = 1 + 2 + 3 = 6$$

and so on.

(i) Write down the value of T_4.
(ii) Find $(T_5 - T_4)$, $(T_{12} - T_{11})$, $(T_{27} - T_{26})$ and write down an expression for $(T_{n+1} - T_n)$ in terms of n.
(iii) Given that $T_{100} = 5050$, find T_{101} and T_{102}.
(iv) Prove or refute the conjucture '$T_{n+2} - T_n$ is always odd.'
(v) Prove, or show by some means that

$$T_n = \frac{n(n+1)}{2}$$

Hence find T_{1000}.

Question 2

The tetrahedral numbers H_1, H_2, H_3 etc. are calculated as follows:

$$H_1 = T_1 = 1 \quad H_2 = T_1 + T_2 = 4 \quad H_3 = T_1 + T_2 + T_3 = 10$$

where T_n is the nth triangular number.

(i) Write down the value of H_4.
(ii) Find $(H_5 - H_4)$ and $(H_{10} - H_9)$.
(iii) Write down an expression for $(H_{n+1} - H_n)$ in terms of the triangular numbers and in terms of n.

(iv) Given that $H_{10} = 220$, find H_{11} and H_{12}.
(v) Prove or refute the conjecture '$H_{n+2} - H_n$ is a perfect square'.
(vi) It is known that, for any value of n,

$$H_n = \frac{n(n+1)(n+2)}{k}$$

where k is a certain positive integer. Use any one of the given values of H to find the value of k and hence find the value of H_{30}.

Question 3

The square numbers S_1, S_2, S_3, etc. are calculated as follows:

$$S_1 = 1^2 = 1 \qquad S_2 = 2^2 = 4 \qquad S_3 = 3^3 = 9$$

and so on.

(i) Write down the value of S_4.
(ii) Find the $(S_5 - S_4)$, $(S_{10} - S_9)$ and write down an expression for $(S_{n+1} - S_n)$ in terms of n.
(iii) Given that $S_{100} = 10\,000$ find S_{101}.
(iv) Prove or refute the conjecture '$S_{n+2} - S_n$ is a multiple of 4'.
(v) Prove, or show by some means, that $S_n = T_{n-1} + T_n$ where T_n is the nth triangular number.
(vi) Comment on this argument: Since $S_n = T_{n-1} + T_n$, it means that every square number is the sum of 2 triangular numbers. Thus you need 2 triangular numbers to make up a square number, so there can only be half as many square numbers as there are triangular numbers.'

Question 4

The pyramid numbers P_1, P_2, P_3, etc. are calculated as follows:

$$P_1 = 1^2 = 1 \quad P_2 = 1^2 + 2^2 = 5 \quad P_3 = 1^2 + 2^2 + 3^2 = 14$$

and so on.

(i) Write down the values of P_4 and P_5.
(ii) Find $(P_5 - P_4)$ and an expression for $P_{n+1} - P_n$ in terms of n.

(iii) Given that $P_{10} = 385$, find P_{11}.
(iv) Find the value of n for which $P_{n+1} - P_n = 169$.
(v) Give a reason why $P_{n+2} - P_n$ is always an odd number.
(vi) It is known that, for any value of n,

$$P_n = \frac{n(n+1)(2n+1)}{k}$$

where k is a certain positive integer. Use any one of the given values of P_n to find the value of k, and hence find the value of P_{24}.

Question 5

Try to give a geometric explanation of the theorem

$$n \times m = T_n + T_m - T_{n-m}$$

where $(n < m)$.

Question 6

Prove or refute the result

$$P_n = 2(T_1 + T_2 + T_3 \ldots + T_{n-1}) + T_n$$

Question 7

The cube numbers are calculated as follows:

$$C_1 = 1^3 = 1 \quad C_2 = 2^3 = 8 \quad C_3 = 3^3 = 27$$

and so on.

Experiment with the sums and differences and try to find out as much information as you can of the type we have looked at in questions 1–6.

There is a result connecting the cube and the triangular numbers. Can you find it?

Question 8

On some large sheets of paper make a wall display illustrating something about triangular, tetrahedral, square, pyramid or cube numbers. If you have the necessary materials make some 3-D models of the tetrahedral, pyramid or cube numbers.

33 Tile Pattern Problem

Suppose you wish to cover a rectangular space which measures 20 × 50 centimetres (cm) with tiles.

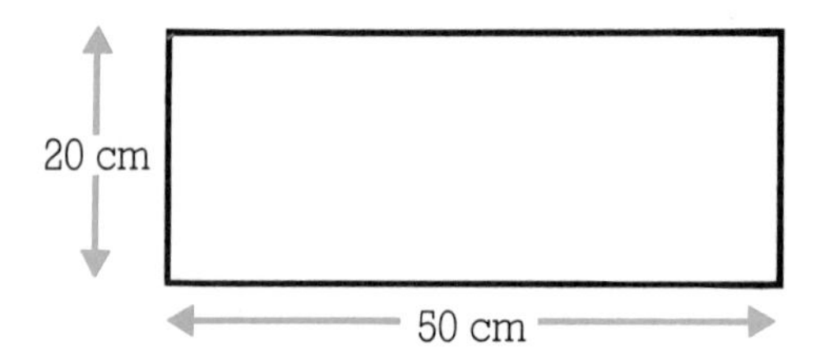

Each tile is also a rectangle, but measures 20 × 10 cm.

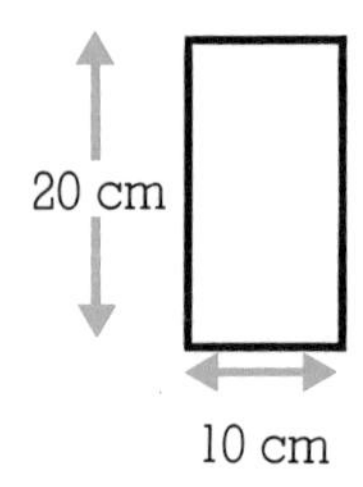

Here are two ways you could lay the tiles:

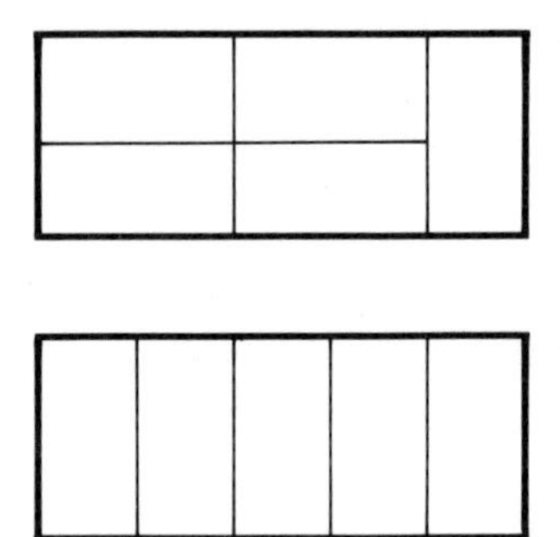

There are 8 different ways of filling the space with these tiles.

Question 1

Can you find all 8 ways of filling the 20 × 50 cm space with 20 × 10 cm tiles? Draw a diagram for each.

Question 2

Using a 20 × 10 cm tile, sketch all the tile patterns you can find for filling the following spaces:

1 20 × 10 cm rectangle
2 20 × 20 cm rectangle
3 20 × 40 cm rectangle

Question 3

Now try to find the number of tile patterns for:

1 20 × 60 cm rectangle
2 20 × 70 cm rectangle

You need *not* sketch all of the patterns.

Question 4

Record all your results in a table or similar. Give an idea of how you tackled the problem. Write down any observations you made. **Generalise** your results. Write a neat report of this work.

Question 5

(i) All of the rectangular spaces in questions 1–4 were:

20 × something cm.

You might like to consider looking at spaces which are:
30 × something cm
40 × something cm
50 × something cm
etc.

(ii) You might wish to change the size of the tile, for instance 20 × 5 cm or 20 × 20 cm.

(iii) You might wish to have different coloured tiles so that, for example, these arrangements are different:

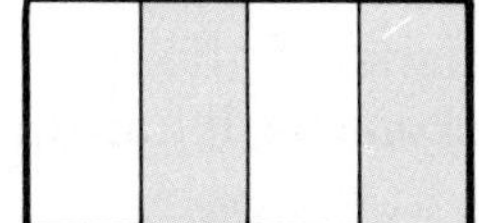

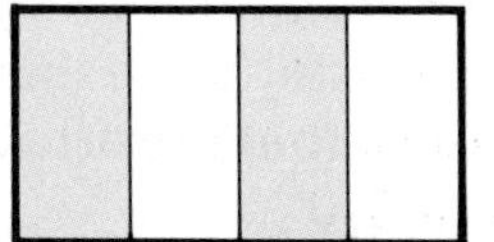

34 Exterior Angles

As well as having **interior** angles, a figure also has **exterior** angles. This diagram shows an interior and exterior angle for a triangle:

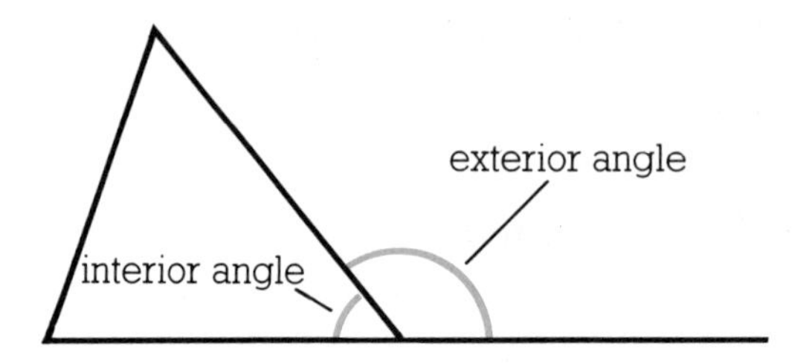

or for a hexagon:

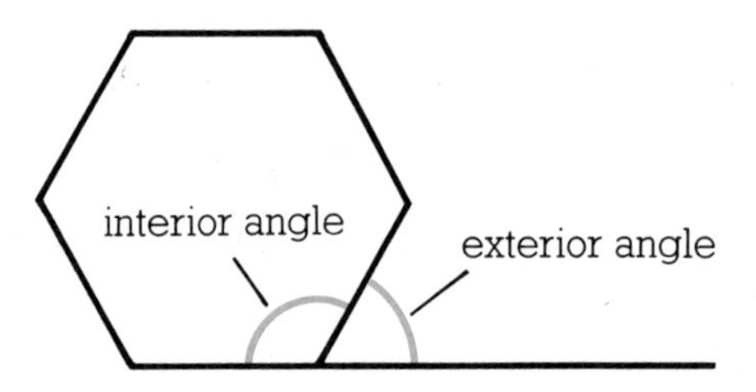

We can examine the exterior angles of a triangle by following an arrow around the sides of the triangle and adding up the angles through which it turns. The arrow *slides* along sides and *turns* — always anti-clockwise — through each of the exterior angles. As it moves completely around the triangle the arrow starts like this: ⟵ and finishes in the same position like this: ⟵ Notice that the arrow can move backwards. It has turned through 4 right angles or 360°.

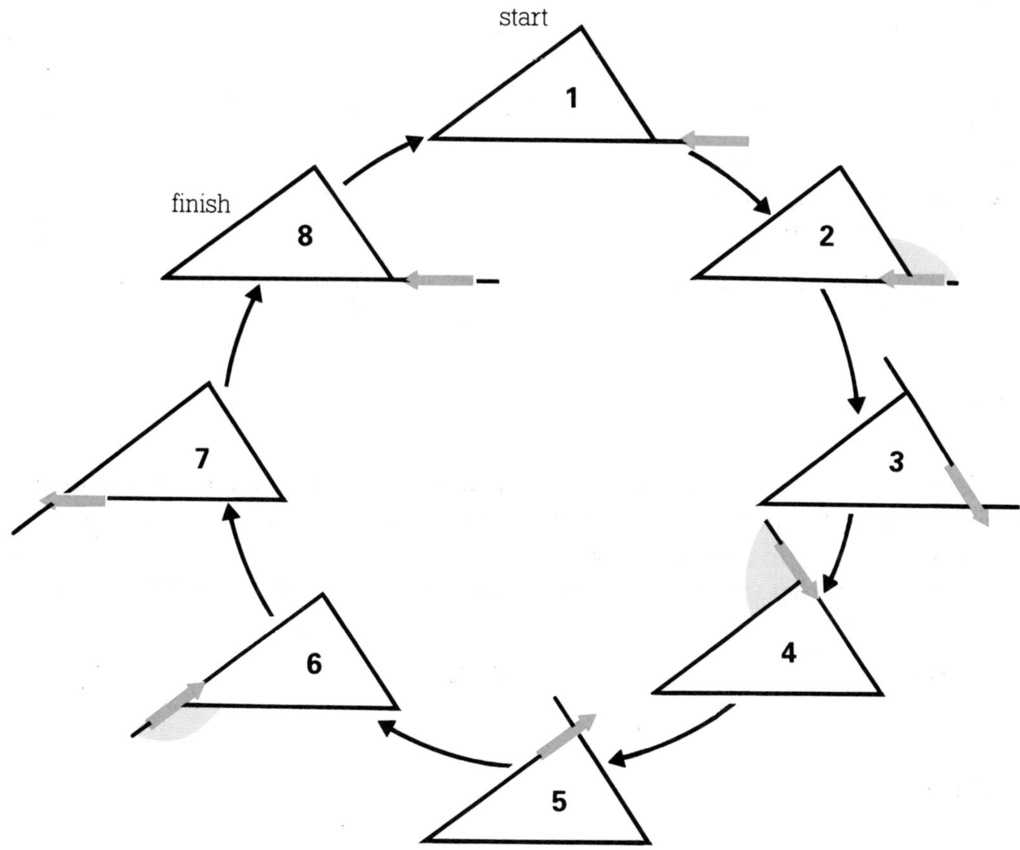

Question 1

Draw 3 triangles of your own. Move your arrow completely around each triangle, sliding it along the sides and turning it — always anti-clockwise — through each exterior angle.

If you get it to turn through something other than 4 right angles in each case, show your teacher. If it turns through 4 right angles in all cases, move on to the next questions.

Question 2

Draw three 4 sided figures. Move your arrow completely around each one, again by sliding it along the sides and turning it — always anti-clockwise — through each exterior angle. How many right angles does it turn through in each case? Your answers should all be the same. Go on to the next question if they are.

If they are different, speak with your teacher.

Question 3

Do what you have done with the triangles and 4 sided figures for:

1 two 5 sided figures
2 two 6 sided figures
3 one 7 sided figure

Count and *record* the number of right angles the arrow turns through in each case.

Question 4

There is something special about the number of right-angles the arrow turns through. Can you spot what this is? Discuss it with your neighbours and teacher and write up any conclusions you come to.

Could you predict the number of right angles the arrow would turn through if the figure had

1 10 sides **2** 23 sides?

Question 5

Calculate the exterior angles of:

1 an equilateral triangle
2 a regular pentagon
3 a regular hexagon
4 a regular polygon which has 20 sides

Question 6

Each diagram shows one of the *exterior* angles of a regular polygon:

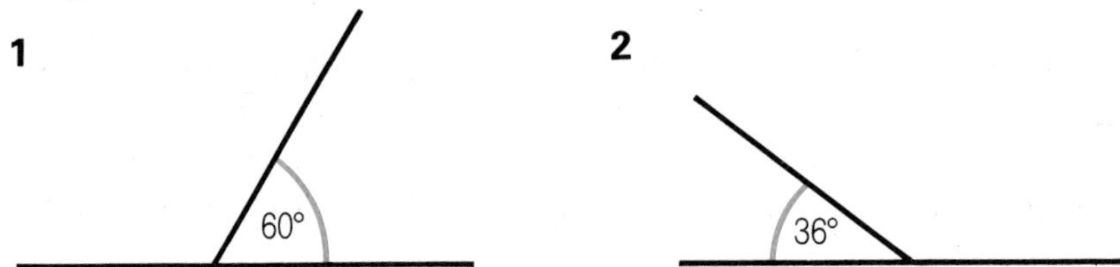

How many sides do each of these polygons have?

Question 7

For a particular regular polygon, the interior angle is 3 times as big as the exterior angle. Calculate the number of sides on that polygon. Show all your work and discuss this with your neighbours.

Question 8

A regular polygon has an exterior angle $x°$ and an interior angle $5x°$. How many sides does the polygon have?

Question 9

A regular polygon has an exterior angle $x°$ and an interior angle $kx°$, where k is a whole number. Prove that the polygon has $2k + 2$ sides.

Question 10

For some regular polygons the interior angle is an integer multiple of the exterior angle; for other polygons this is not the case. What is it that separates these two possibilities? Explain your answer.

35 Congruent Parts

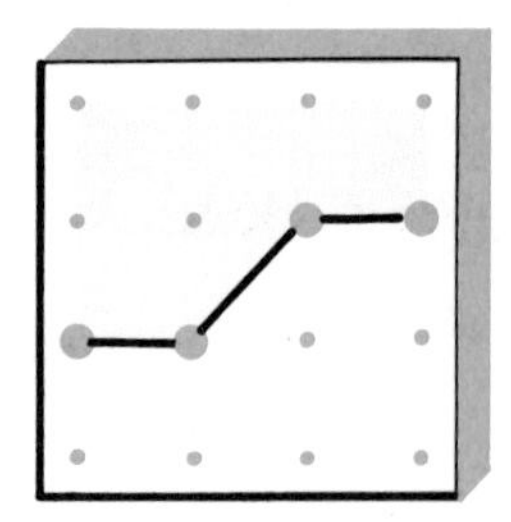

This 16 pin geoboard has been divided into 2 **congruent** parts. What happens to the 2 parts when the board is turned half a complete turn?

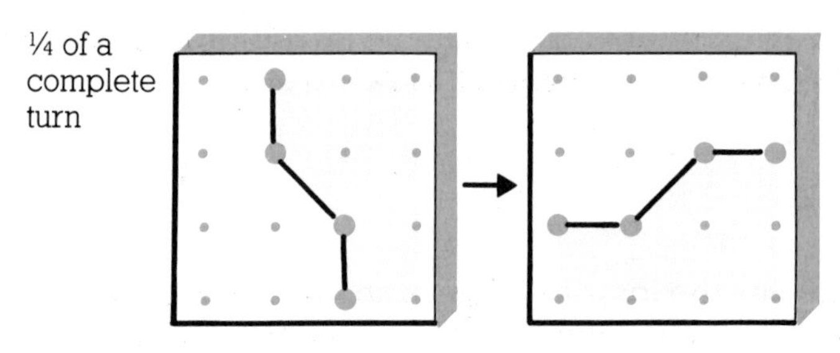

Compare the shapes on the board before and after turning. What does **congruent** mean?

Question 1

How many different ways are there of dividing the board into 2 congruent parts?

Record and draw all your results.

Try to find a method for working out the different ways, so that you know when you have found all the possibilities.

Question 2

How many ways are there of dividing the board into 4 congruent parts?

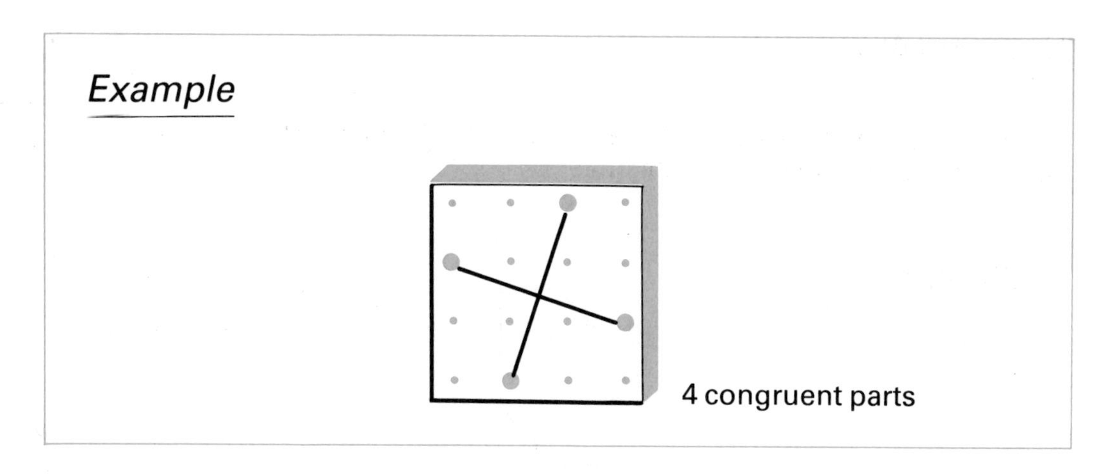

Question 3

Try dividing up the 9 pin and 25 pin boards into different numbers of congruent parts.

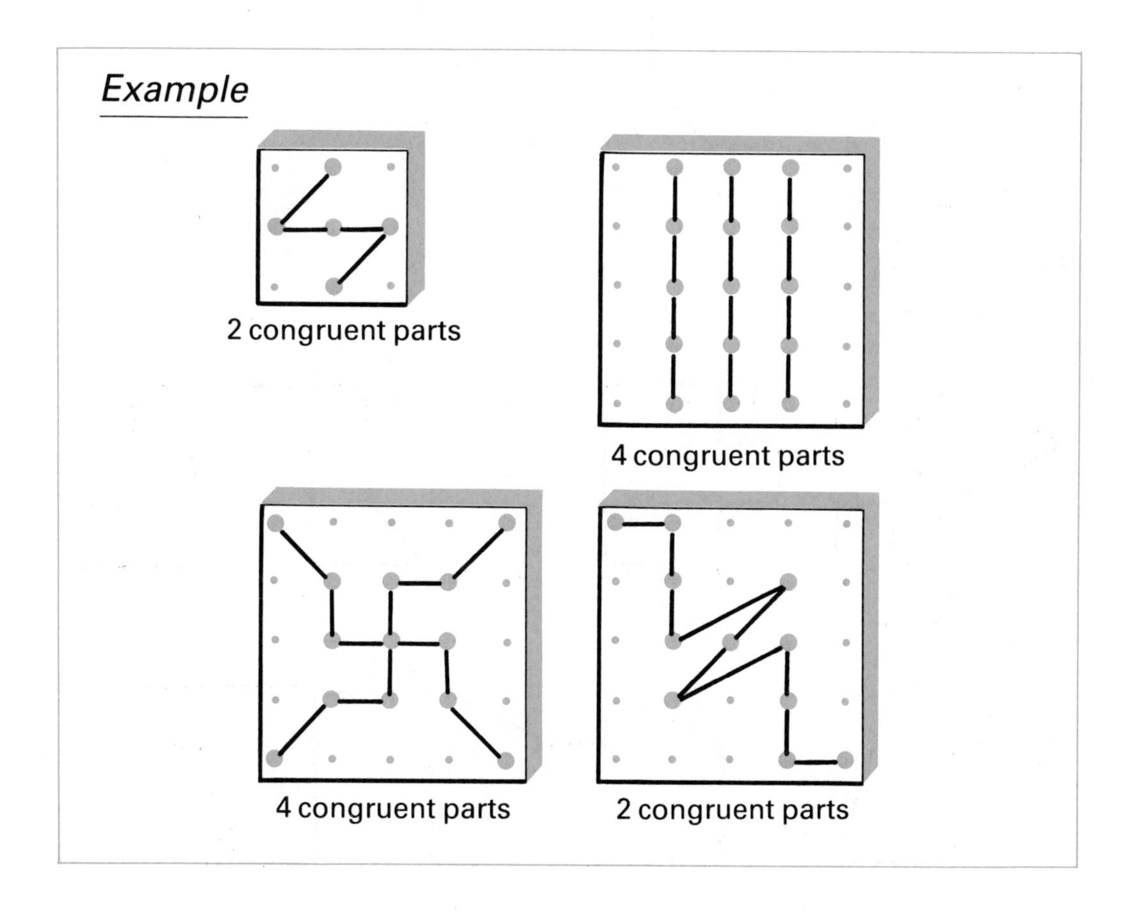

Question 4

In the following, try to separate the shapes into 2 congruent halves — not necessarily with a single straight line.

Example

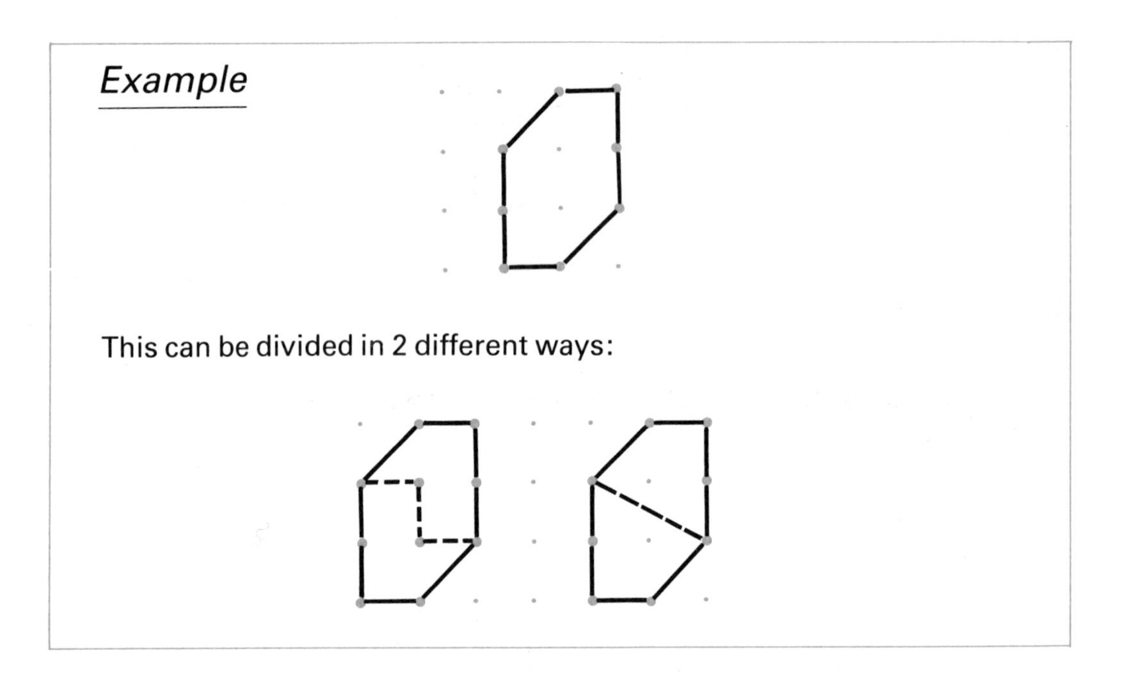

This can be divided in 2 different ways:

Now try these:

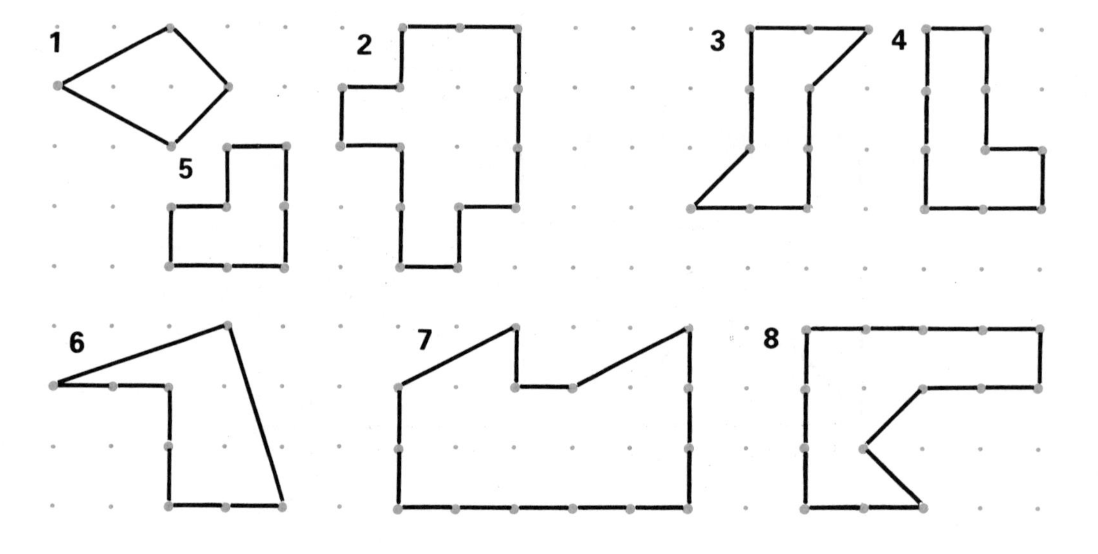

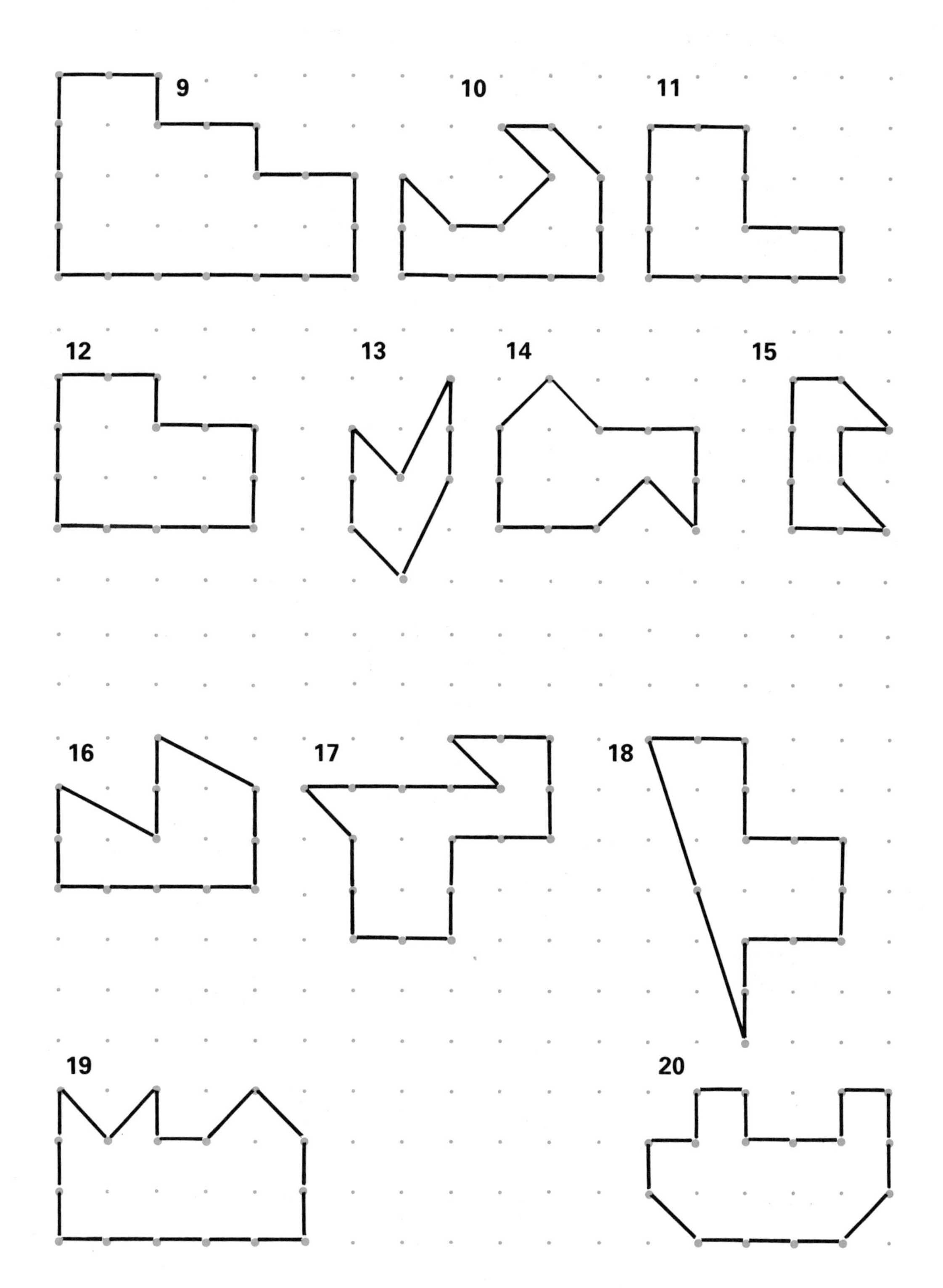
9
10
11
12
13
14
15
16
17
18
19
20

The following shapes are all squares. They have the same shape but are different sizes and are called **similar**.

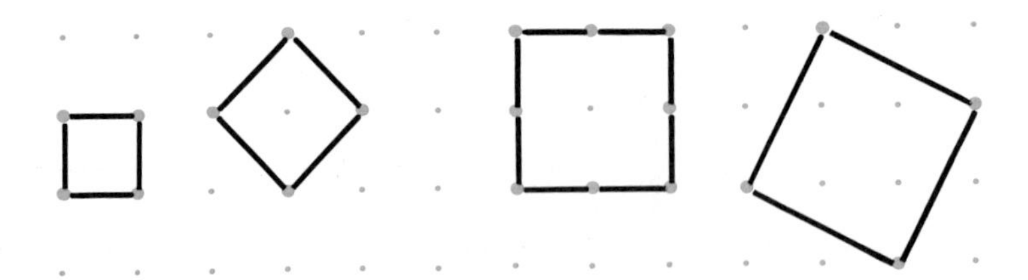

Question 5

On the geoboard, or on dotted paper, use the original shapes shown below to produce similar shapes.

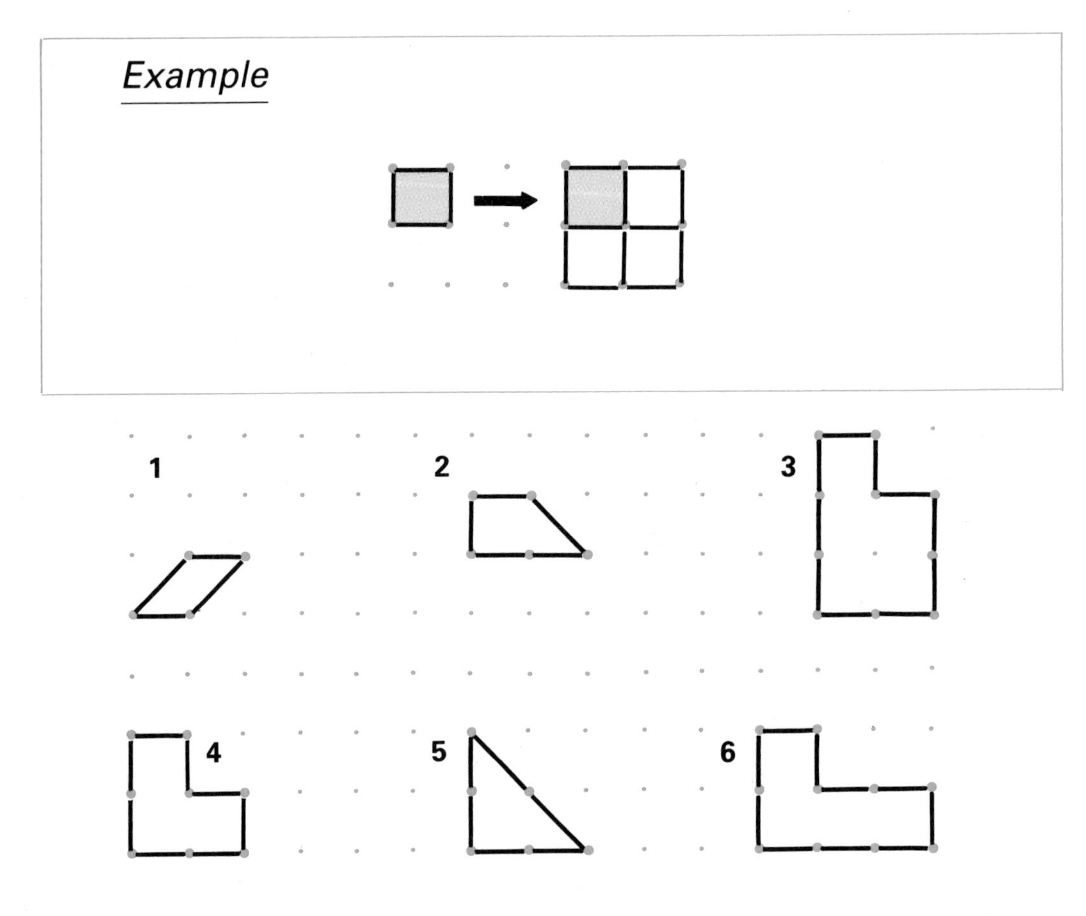

36 Shuffles

A **shuffle** is a rearrangement of a pile of cards following a precise set of rules. Here is an example of one particular shuffle.

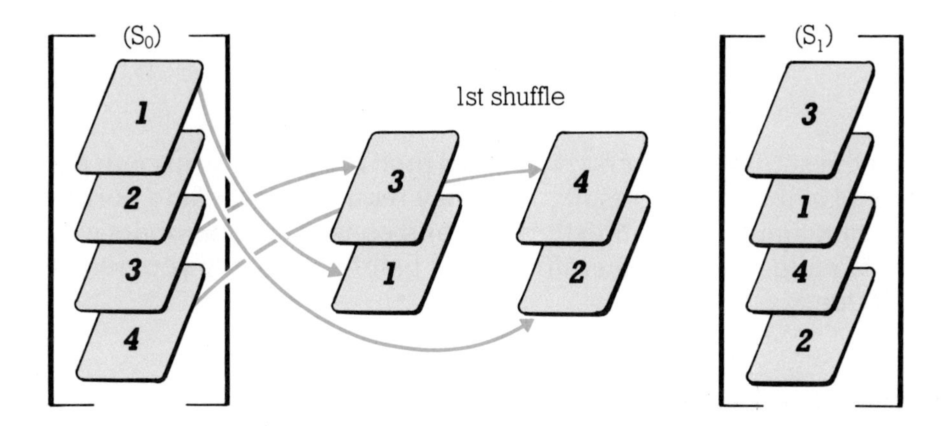

Here, the cards (1, 2, 3, 4) are shuffled by placing them alternately in two piles and then placing the left-hand pile on the right. This completed shuffle (S_1) produces the new card-order of 3, 1, 4, 2.

You can see that the next shuffle (S_2) gives the next card-order of 4, 3, 2, 1.

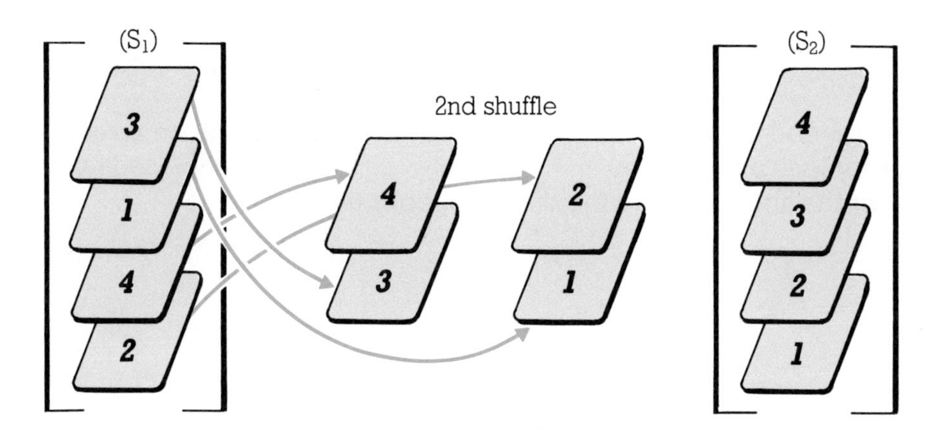

You might like to check that in this case the result of the 4th shuffle (S_4) brings the cards back to their original order (S_0). So that, for this shuffle:

$$(S_4) = (S_0)$$

Question 1

Choosing 6 cards, and using the Joker or a picture card as the zero, invent precise rules for your own shuffle. After trying first to visualise and predict the results of each shuffle, record the results like this:

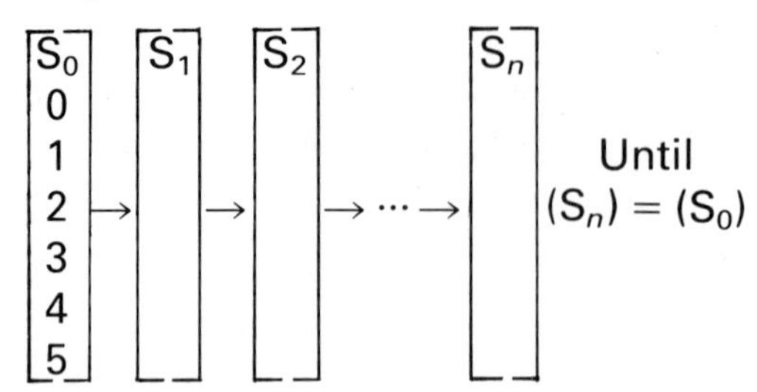

Explain the shuffle in your own way and then identify any number patterns you can find in the table of results. This may require you looking horizontally, vertically or diagonally through the table and considering numbers in different modulo. You may therefore find it helpful to record your results on squared paper.

Arrow diagram

On the shuffle in the introduction, we saw that:

$$\begin{bmatrix} S_0 \\ 1 \\ 2 \\ 3 \\ 4 \end{bmatrix} \text{ became } \begin{bmatrix} S_1 \\ 3 \\ 1 \\ 4 \\ 2 \end{bmatrix}$$

This information could have been represented in the form of an **arrow diagram** for the cards where ——▶—— means 'is replaced by'
So that for the shuffle above the diagram looks like this:

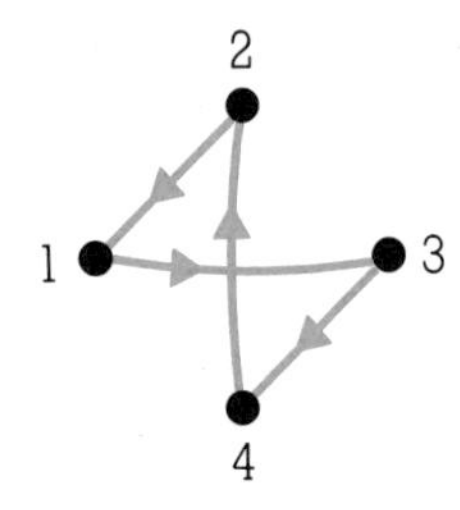

This could be simplified by changing the position of the card-spots to:

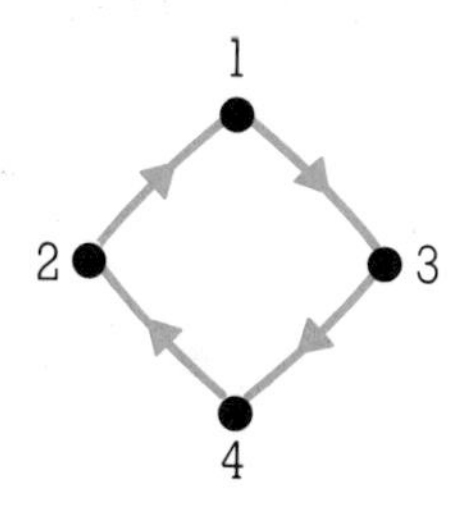

The diagram for the next part of the shuffle is shown below:

S_1	S_2
3	4
1	3
4	2
2	1

This gives an arrow diagram of:

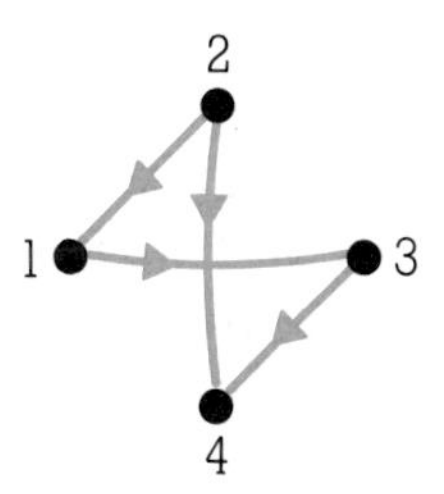

Now check the diagrams for (S_2) to (S_3) and (S_3) to (S_4) and write down anything you notice about them.

Question 2

Use an arrow diagram or some other efficient method to work out when the shuffles below return to their original order. Check by actually performing the shuffle where possible.

(i) Deal 6 cards alternatively into 2 piles, the first pile being on the left. Then put the left pile *on* the right pile.

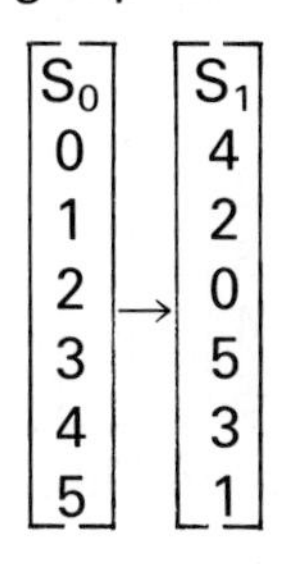

S_0		S_1
0		4
1		2
2	→	0
3		5
4		3
5		1

(ii) Deal 4 cards into 3 piles, the first pile being on the left. Then put the middle pile on the left pile and then these on the right pile.

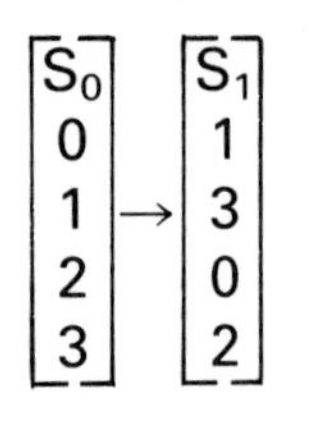

S_0		S_1
0		1
1	→	3
2		0
3		2

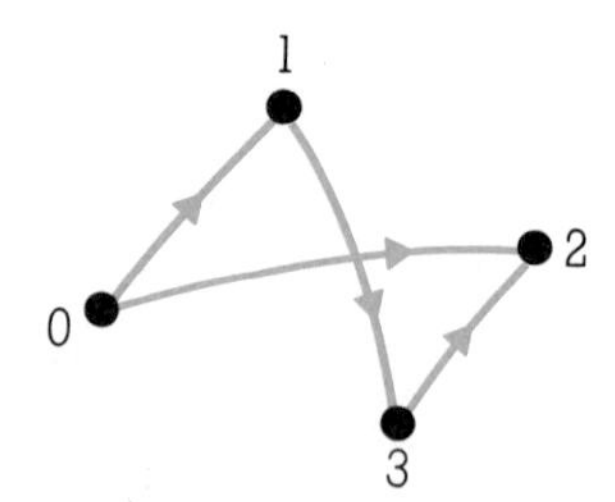

(iii) Deal 8 cards into 3 piles, the first in the middle, the second on the left, the third on the right. Then put the left pile on the middle and these on the right pile.

S_0		S_1
0		7
1		4
2		1
3	→	6
4		3
5		0
6		5
7		2

Draw two different arrow diagrams for this shuffle.

Question 3

Being as adventurous as possible, work out, then describe, some shuffles which produce the following:

1

S_0		S_1
0		3
1		0
2	→	4
3		1
4		5
5		2

2

S_0		S_1
0		5
1		2
2	→	1
3		4
4		3
5		0

3

S_0		S_1		S_2
0				2
1				5
2	→	?	→	1
3				0
4				4
5				3

4

S_0		S_1		S_2
0				3
1				1
2	→	?	→	4
3				2
4				5
5				0

Try to use shuffles with only 2 or 3 different piles in each case.

Question 4

(Harder)
If you choose 2 different shuffles and *alternate* them, can you predict the total *number* of shuffles required to restore the cards to their original order? For example, the first and then the second shuffle in question 2.

Question 5

(Much harder)
For the first shuffle in question 2
with 2 cards $(S_0) = (S_1)$
with 4 cards $(S_0) = (S_4)$
with 6 cards $(S_0) = (S_6)$
with 8 cards $(S_0) = (S_3)$
etc

We see that the number of cards in the pack affects the number required for the pack to return to its original order.

For *n* cards in the pack, can you find (in terms of *n*) the number of shuffles required for the shuffle to return to S_0, i.e.

$$S_{f(n)} = (S_0)?$$

37 Cinema Problem

The manager of a cinema has his own system of prices:

adults	£10
pensioners	50p
children	10p

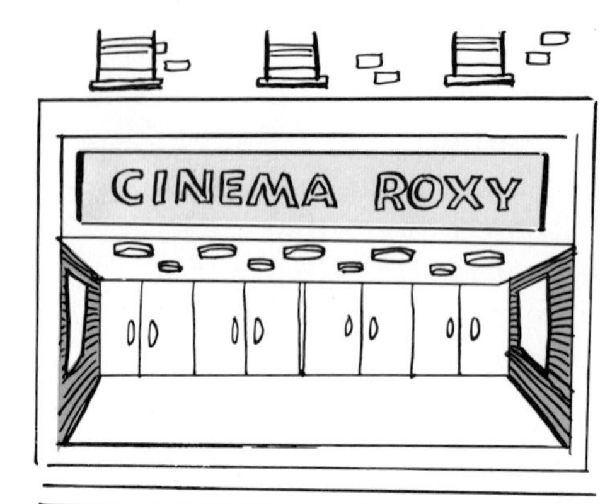

There are exactly 100 seats.

If there were 10 adults, 40 pensioners and 50 children the cinema would take:

$$(10 \times £10) + (40 \times 50p) + (50 \times 10p)$$
$$= £100 + £20 + £5$$
$$= £125$$

Question 1

How much would the cinema take if they had:

1 6 adults, 50 pensioners and 30 children
2 12 adults, 80 pensioners and 8 children.

Question 2

The manager wonders if he can fill exactly 100 seats and take exactly £100. Can you show him how to do this?

Question 3

Can he fill exactly 100 seats and take exactly £100 if his prices are:

1 adults £10
pensioners £1
children 50p

2 adults £10
pensioners 50p
children 2 for 25p?

Question 4

Modify either the number of seats or the prices to create a similar problem for your neighbours.

38 Cuisenaire Activity 5
Rod equations

w : *white*
r : *red*
g : *light green*
p : *pink*
y : *yellow*
d : *dark green*
b : *black*
t : *tan* (brown)
B : *blue*
o : *orange*

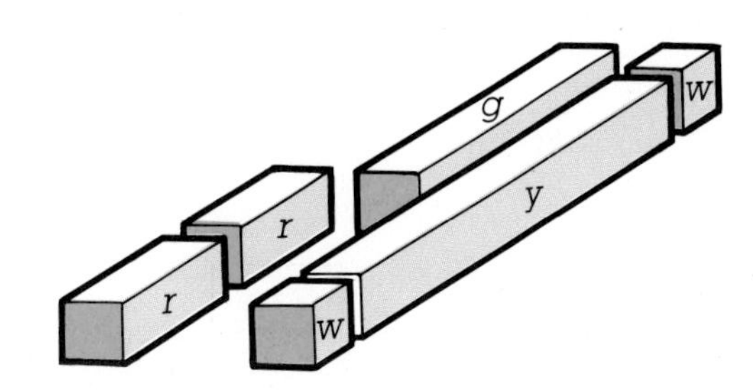

Rod equations

There are many ways of representing the rod arrangement shown below:

y		
r	*r*	*w*

These are some:

$$y = r + r + w$$
$$y = 2r + w$$
$$y = w + 2r$$

Still using only these rods we could also have . . . . .

$$y - r = r + w$$

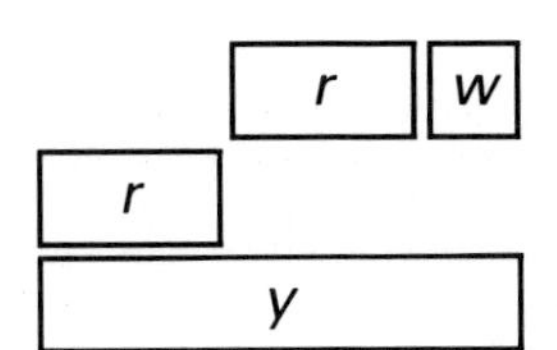

and similarly:

$$2r = y - w \text{ and } r = \tfrac{1}{2}(y - w)$$

Try to produce equations which cannot be physically constructed even though they use the rods shown.

Example

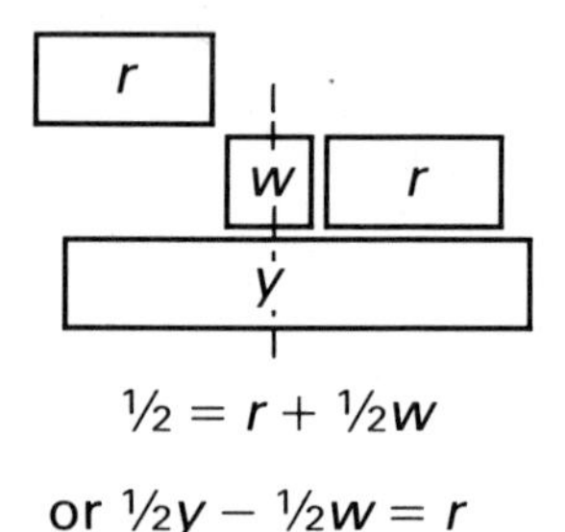

$$\tfrac{1}{2} = r + \tfrac{1}{2}w$$

$$\text{or } \tfrac{1}{2}y - \tfrac{1}{2}w = r$$

This could be compared with similar results using these rods such as:

$$\tfrac{1}{2}(y - w) = r$$

or

$$y - 2(2r + w) = -r - r - w$$

Can you think of others?

Question 1

Give rod statements for the following arrangements:

1

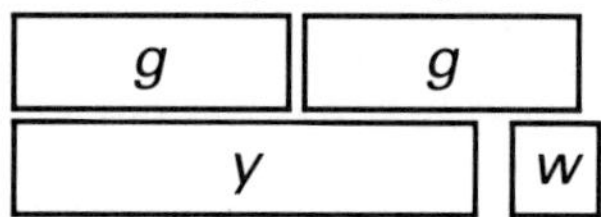

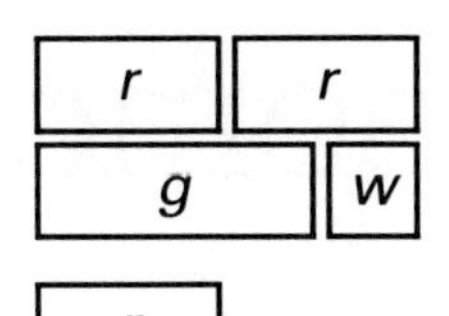

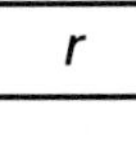

r

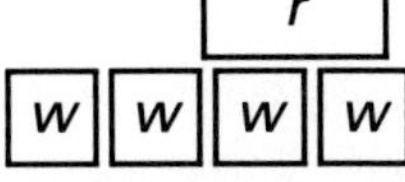

4

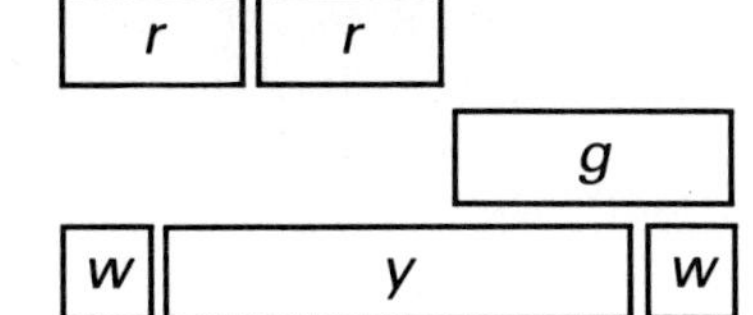

Question 2

Sketch the rod arrangements from the following statements:

1 $3r = p + r$
2 $r + 2g = b + w$
3 $y - w = p$
4 $w = \frac{1}{2}r$
5 $w = \frac{1}{3}(B - d)$
6 $p = \frac{1}{2}d + w$

Question 3

Using only the rod indicated, produce as many different arrangements and statements of the following as you can. Illustrate some of your answers with quick free-hand sketches.

1 $y = w + w + w + w + r$
2 $b = g + g + w$
3 $B = r + r + r + r + w$
4 $w + w + r + r = d$
5 $B = r + r + p + w$
6 $b + 2p = 4g + 3w$

If you attempt some more of your own, move as quickly as possible to finding complicated statements.

39 Trigonometry 1
Sines and cosines

Many years ago, people were faced with certain problems that seemed very different but were actually related. Some were practical problems, others not. One of the practical problems was measuring the time of day. This could be estimated according to the length of shadows cast.

This is near mid-day when shadows are short.

This is early evening when shadows are long.

This was a very practical problem. A second one was much less use but appealed to the curious. It was about the relationship between the length of a **chord** of a circle and the angle that chord makes at the centre of the circle.

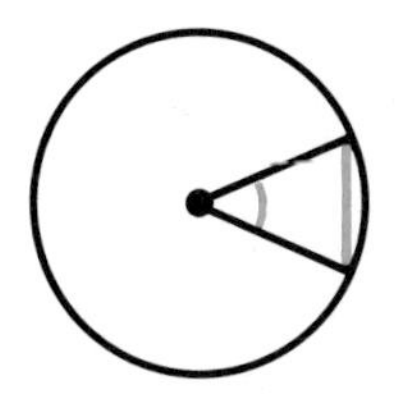

A fairly short chord makes a smallish angle.

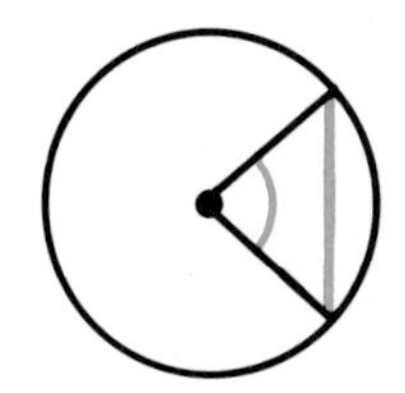

A longer chord makes a much larger angle.

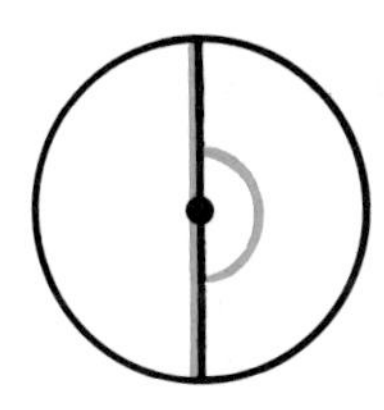

The longest possible chord makes an angle of 180°.

These 2 problems are connected by a topic we now call **trigonometry**.

Question 1

To attempt the problem of finding a relationship between chords, lengths and angles inside a circle, use one of the methods suggested below or think of one yourself. In each case you will need a protractor.

(i) Draw circles of about 10 cm radius and draw chords of different lengths.

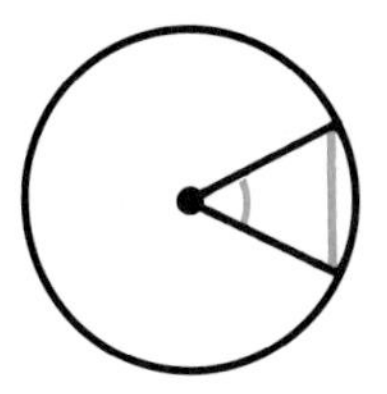
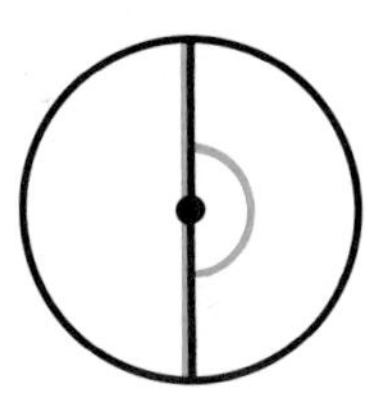
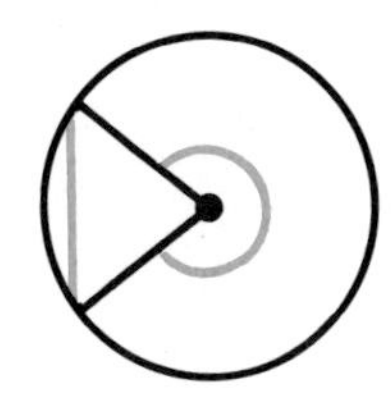

(ii) Use geostrips and a metre ruler.

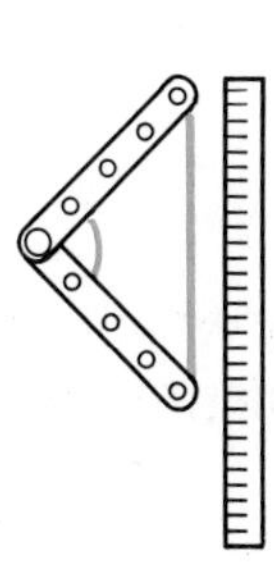

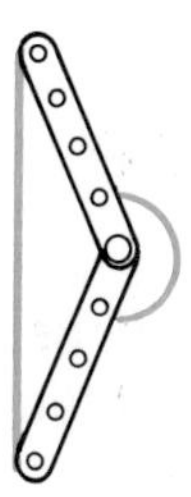

(iii) Use 2 metre rulers and a tape measure.

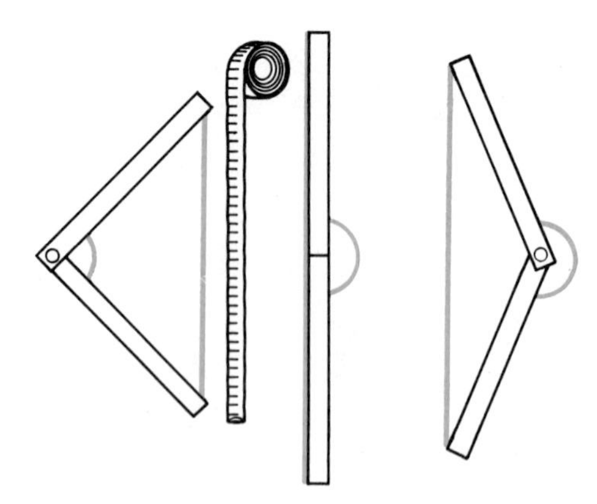

Whichever way you choose, construct a table of results like this:

angle	*chord length*

Choose a wide range of angles from 0° to about 360°. You will probably need around 20 different angles.

Question 2

Draw a graph like this and plot your results on it from question 1.

Sine and cosine

Some special words are used in trigonometry which you will need to learn. These words are illustrated in the following diagrams.

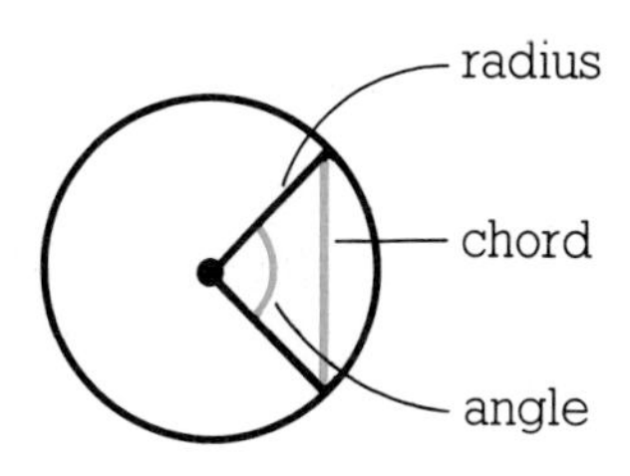

Using half the circle.

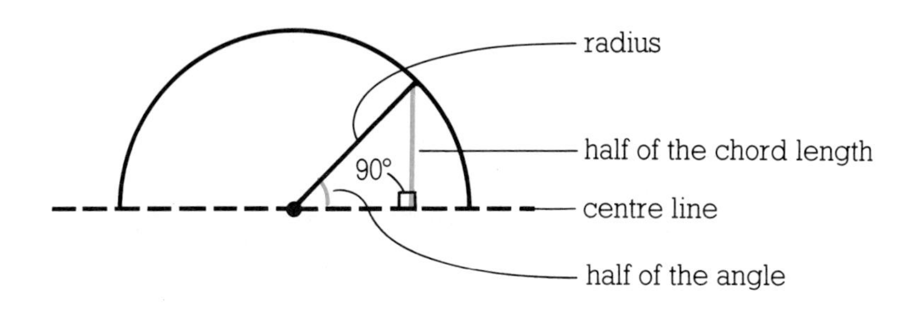

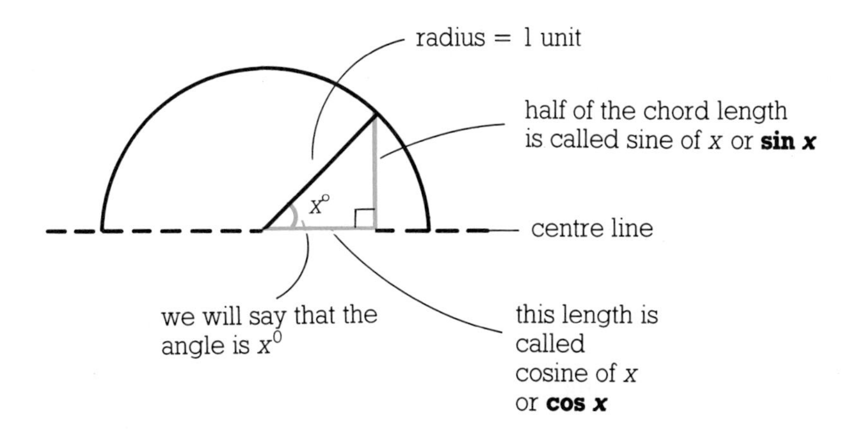

Now the investigation can be developed further. You will need to construct some circles all of radius 1 unit. What you have as 1 unit is your choice.

You might like:

1 metre	This makes for big circles.
10 centimetres	This is exactly the length of 1 *orange* cuisenaire rod.
or 1 geostrip	You will have to work out fractions of a geostrip.

Whatever you choose, draw a few semi-circles such as:

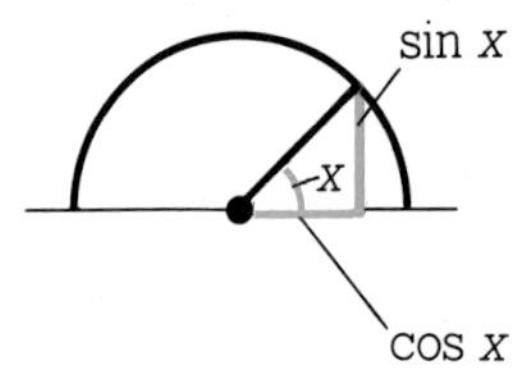

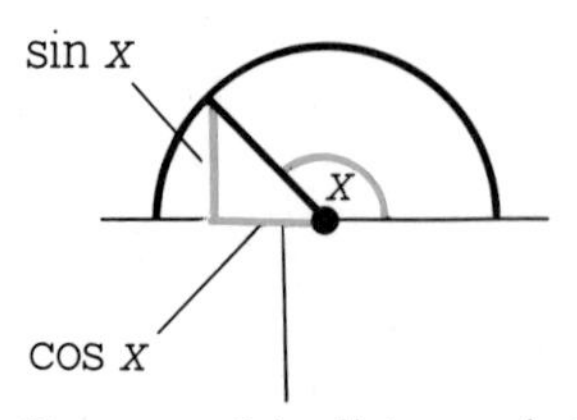

Because this distance is to the left of centre we regard it as a negative value.

Question 3

For about ten values of the angle $x°$, between 0° and 180°, with the radius of 1, measure in each case:

the angle x
the value called sin x
the value called cos x.

Set up a table of values like this:

angle $x°$	*sin x*	*cos x*

Then draw graphs like the ones shown and plot your results on them.

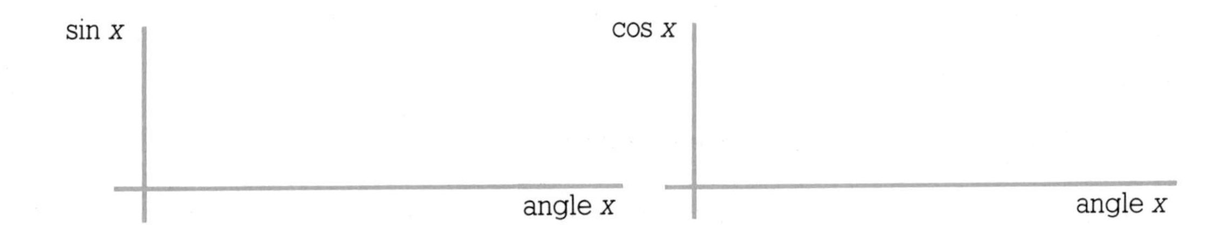

Using a calculator

If you have a scientific calculator, then you can check on the accuracy of your results. You can also check by using a book of mathematical tables which has sine and cosine values in it.

For instance, from a scientific calculator or tables, you should be able to find that:

$$\left.\begin{aligned} \sin 75^\circ &= 0.9659 \\ \cos 30^\circ &= 0.8660 \end{aligned}\right\} \text{to 4 decimal places}$$

Using the calculator to get sin 75°, what you must do is:

(i) make sure your calculator is working in degrees
(ii) key in 75
(iii) press the button marked 'sin'.

If your results are not very accurate, make a new table and fill in the accurate ones. Then redraw the 2 graphs.

Question 4

Use a scientific calculator to obtain the values of:

1 sin 130°
2 cos 130°
3 sin 105°
4 cos 120°
5 sin 120°
6 cos 120°

Question 5

Find all of the angles between 0° and 180° for which:

1 $\sin x = 0.5$ **2** $\sin x = \cos x$

Question 6

So far we have only looked at angles between 0° and 180°. Investigate sines and cosines for angles between 180° and 360°. Draw pictures and graphs, and construct tables of values.

40 Trigonometry 2
Tangent investigation

We shall start by going back to the pictures of a semi-circle from Trigonometry 1.

To add a tangent to our drawing we draw a line perpendicular to the radius.

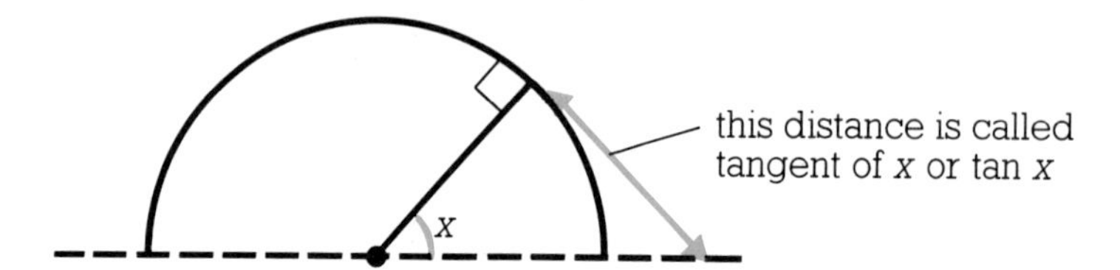

The tangent in this diagram is to the left of centre. We therefore say its length has a negative value.

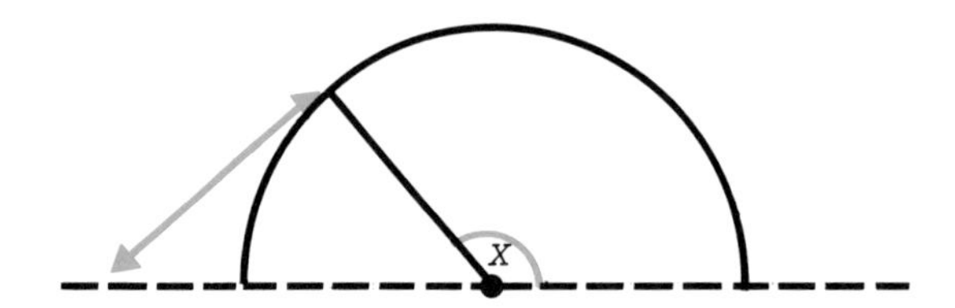

Question 1

Draw some semi-circles all of radius 1 unit. What you have as 1 unit is your choice.

Choose a range of angles between 0° and 180° — about 15 angles in all. In each case draw and measure the length of the tangent. Record your results like this (some should be negative):

angle	*tangent*

You can find the accurate tangent values using a scientific calculator (or mathematical tables for angles from 0° to 90°). For instance:

$$\tan 50° = 1.1917$$

To do this calculation you must

(i) key in 50
(ii) press the button marked 'tan'.

Question 2

For the angles you used in question 1, make another table like this:

angle	*tangent by calculator*

Check on the accuracy of your results in question 1.

Using your table draw the graph:

You will need to be very careful about tan 90°. What do you think its value is?

Using a scientific calculator, we shall do a small calculation. We choose any angle to do it, say 25°.

$\sin 25° = 0.4226183$
$\cos 25° = 0.9063078$
$\sin 25° \div \cos 25° = 0.4226183 \div 0.9063078$
$= 0.4663077$
$\tan 25° = 0.4663077$

So we have the result:

$\sin 25° \div \cos 25° = \tan 25°$

Question 3

Choose any four angles of your own. You should be able to show that for each different angle:

$$\text{sin angle} \div \text{cos angle} = \text{tan angle}$$

$$\text{or } \frac{\sin x^\circ}{\cos x^\circ} = \tan x^\circ$$

Question 4

By any method you can, solve each of these equations:

1 $\tan x = 1$
2 $\tan x = 0.3$
3 $\tan x = -1$
4 $\tan x = -0.5$
5 $\tan x = 100$
6 $\tan x = 0.5$
7 $\tan x = 2.0$
8 $\tan x = -5$
9 $\tan x = 5$
10 $\tan x = -100$

Question 5

Explain why

$$\tan 90^\circ = \infty \text{ (infinity)}$$

Question 6

By any method you can, solve the equation:

$$\tan x = \tan 2x$$

Question 7

Investigate with illustrations and values the meaning of *tangent* for angles between 180° and 360°.

41 Number Theory 2
Modulo arithmetic

The work on last digits in the Chapter 31, Number Theory 2, is usually called **arithmetic modulo 10**. In this arithmetic we only have the digits

0, 1, 2, 3, 4, 5, 6, 7, 8, 9,

and we construct rules for operating with these numbers based on a sort of clockface. That is, we set the numbers up as:

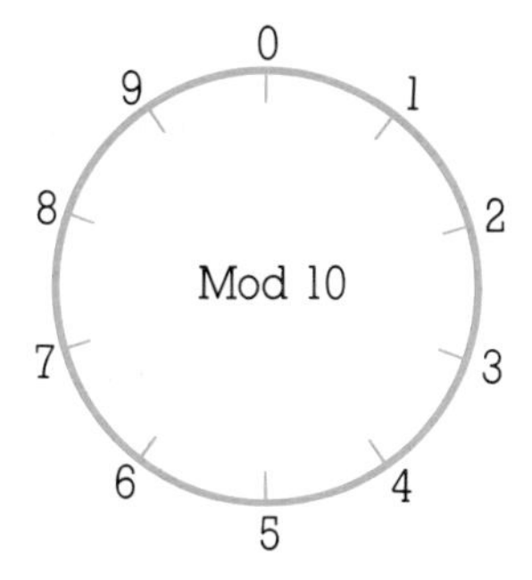

Then to do say, 5 + 8 in this arithmetic, we start at 5 and go around the clock face, or circle, clockwise, counting on another 8. When we have done this we end up at 3. So we write

$$5 + 8 = 3$$

which looks wrong but we remind ourselves that we are working in modulo 10 arithmetic by writing

$$5 + 8 = 3 \text{ (Mod 10)}$$

Try to get the feeling for modulo arithmetic by attempting these exercises.

Question 1

1 $4 + 7$ (Mod 10)
2 6×3 (Mod 10)
3 5×4 (Mod 10)
4 $7 + 8$ (Mod 10)
5 9×3 (Mod 10)
6 4^2 (Mod 10)
7 7^2 (Mod 10)
8 2^4 (Mod 10)
9 $8 + 7 + 3$ (Mod 10)
10 $6 \times 2 \times 3$ (Mod 10)

We can do the arithmetic in other modulo. For instance in Mod 6, we have the digits

$$0, 1, 2, 3, 4, 5,$$

and the clockface

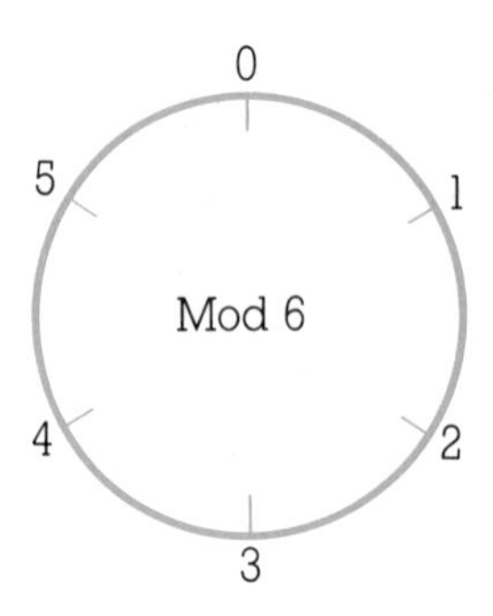

So to calculate $2 + 5$, start at 2 and count on 5 more, going clockwise again. We end up at 1. Therefore,

$$2 + 5 = 1 \text{ (Mod 6)}$$

Now try these, constructing the appropriate clockfaces if you wish.

Question 2

1 $4 + 5$ (Mod 6)
2 3×5 (Mod 6)
3 $2 + 3$ (Mod 6)
4 5×4 (Mod 6)
5 $4 + 5$ (Mod 7)
6 $4 + 5$ (Mod 8)
7 7×3 (Mod 8)
8 5×6 (Mod 9)
9 2×8 (Mod 9)
10 3×2 (Mod 5)
11 4×3 (Mod 5)
12 4×3 (Mod 6)
13 2^3 (Mod 7)
14 3^2 (Mod 5)

You may not need to use the clockface for the next problem.

Question 3

When you do these questions, explain how you do them.

1 $3 + 5$ (Mod 7)
2 5×7 (Mod 8)
3 $3 \times 7 \times 2$ (Mod 9)
4 2^5 (Mod 7)
5 5^2 (Mod 6)
6 $4 + 3 + 2 + 1$ (Mod 5)
7 6×7 (Mod 9)
8 4×8 (Mod 10)
9 $7 \times 4 \times 2$ (Mod 9)
10 4^3 (Mod 5)

Negative numbers

It is possible to give negative numbers an interpretation in modulo arithmetic. How do you think we would represent -2 in Mod 7? Think of the Mod 7 clockface.

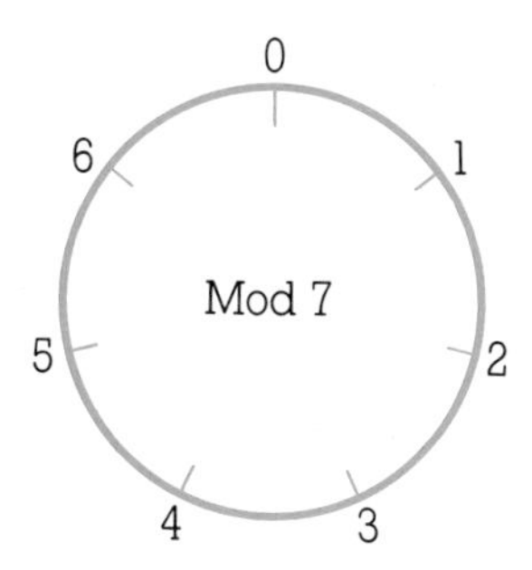

What do you think the answer is? Discuss it with your neighbours and ask for help if you need it.

Question 4

Try to find the answer to these problems without drawing the clockface.

1 -3 (Mod 5)
2 -2 (Mod 9)
3 -5 (Mod 8)
4 -5 (Mod 7)
5 $-2 - 3$ (Mod 6)
6 -2×3 (Mod 8)

42 Geoboard 4

Pick's Theorem

Inside and outside

Shapes on dotted paper can be described in terms of:

(i) (the number of pins on the inside of the shape = I)
(ii) (the number of pins on the boundary of the shape = B).

The shape can then be represented as (I, B)

Example

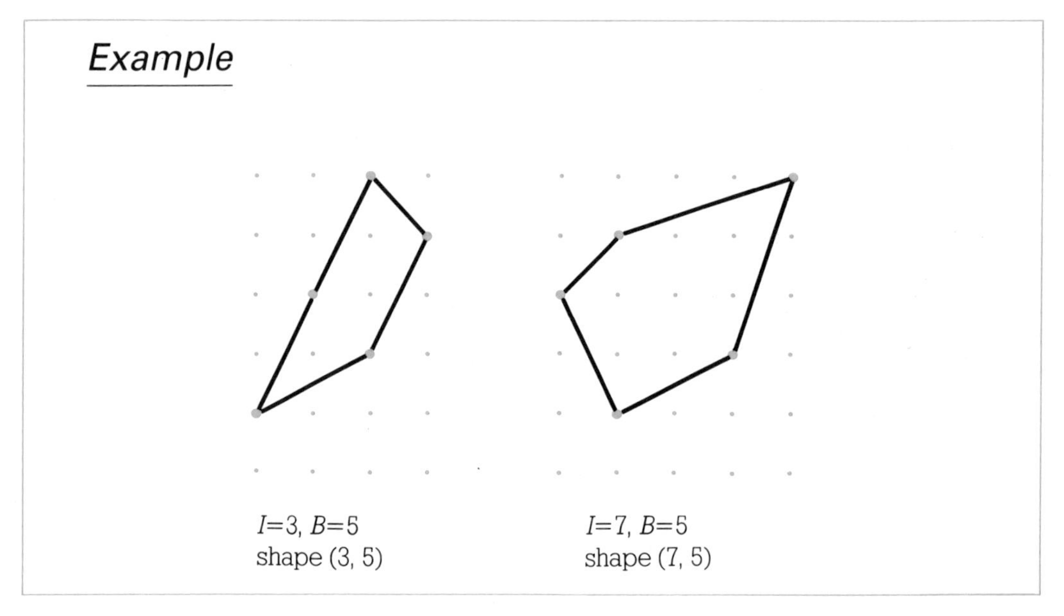

Question 1

How many different shapes for the 2 values of (I, B) can you draw?

Question 2

Are there any values of (I, B) for which shapes cannot be drawn?

Pick's Theorem

Question 3

Find the areas of some of your shapes and draw up a table like the one shown.

	number of pins inside *(I)*	number of pins on boundary (B)	*area of shape*
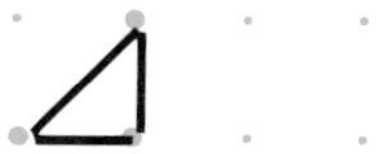	0	3	½
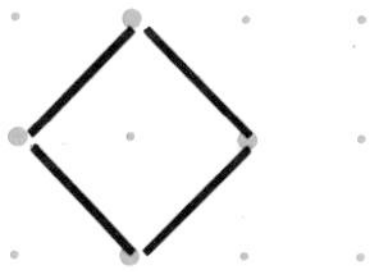	1	4	2
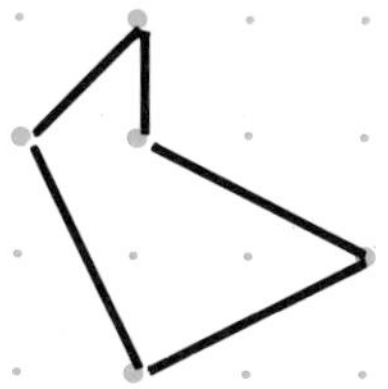 etc.	2	5	3½

Use your table to find out how it is possible to work out the area of a shape given (I, B). Explain clearly how you have set about the task.

The result, when you have found it, is known as Pick's Theorem.

43 Money
A variety of change

If a shopkeeper gave you 25p in change and asked you to guess the values of the 13 coins he was giving you, it might be possible to proceed as follows:
(We will assume that there are no ½ pence or 20 pence coins included.)

Make a guess! $(5 \times 1p) + (8 \times 2p) \rightarrow$ 13 coins but only 21p
Therefore increase the amount:
$(3 \times 1p) + (10 \times 2p) \rightarrow$ 13 coins but only 23p
This is the right direction but increase the amount again:
$(1 \times 1p) + (12 \times 2p) \rightarrow$ 13 coins and 25p!

But this is not the only answer:

If you start with a 5p coin it leaves 20p in change and 12 coins. $(4 \times 1p) + (8 \times 2p) + (1 \times 5p)$ works

Starting with $2 \times 5p$ coins leaves 15p in change and 11 coins. $(7 \times 1p) + (4 \times 2p) + (2 \times 5p)$ also works

There are 5 solutions in total. Can you find the remaining 2 solutions?

Question 1

Find as many ways as possible of making 15p using the coins 1p, 2p, 5p and 10p. Record your results in a table as shown below.

1p	*2p*	*5p*	*10p*	*number of coins*
5	5	0	0	10
3	1	2	0	6
1	2	0	1	4
etc				

Question 2

Can you find a systematic method for finding the 40 ways of making up the value 20p using the coins in question 1?

Question 3

Still using the same coins, find as many of the solutions to the following as you can:

number of coins	*amount of money (in pence)*	*number of solutions*
2	10	1
4	10	1
6	10	2
3	15	1
5	15	2
12	15	1
8	16	3
9	18	4
12	20	3
16	24	3
12	36	7
13	25	5

You can use the computer program which follows (written in BASIC) to check the answers to these and any other questions you attempt. Can you write a different program which also works?

```
 10 LET C=0
 20 PRINT "Number of coins?"
 30 INPUT N
 40 PRINT "Total amount of money?"
 50 INPUT A
 60 PRINT
 70 PRINT;"          1p     2p     5p     10p  "
 80 PRINT,"     ------------------------------
 90 FOR P=0 TO N
100 FOR T=0 TO N-P
110 FOR F=0 TO N-(P+T)
120 FOR E=0 TO N-(P+T+F)
130 IF P+T+F+E=N THEN IF P+2*T+5*F+10*E =A THEN C=C+1 : PRINT
    TAB(1); C;TAB(2);".";TAB(16) ;P TAB(23);T TAB(30) ;F TAB(37) ;E
140 NEXT E
150 NEXT F
```

```
160 NEXT T
170 NEXT P
180 PRINT "
190 PRINT
200 PRINT "The total number of solutions is";C
210 END
```

Here is a typical print out of the program:

```
Number of coins?
?10
Total amount of money?
?24

        1p    2p    5p    10p
      ----------------------
1.      2     6     2     0
2.      4     5     0     1
3.      5     2     3     0
4.      7     1     1     1

All finished

The total number of solutions is 4
```

Question 4

After bringing back the ½p coin and also using the 20p coin, try some of the following (very difficult) questions:

number of coins	*amount of money (in pence)*	*number of solutions*
10	24	4
8	25	4
12	65	21
12	36	9
8	24	4
6	18	3
8	40	8
5	42	3
10	30	6
10	36	8

The program which follows will enable you to check some of the solutions of these and any further questions you attempt.

```
 10 LET Z=0
 20 PRINT "Number of coins?"
 30 INPUT K
 40 PRINT "Total amount of money?"
 50 INPUT J
 60 PRINT
 70 PRINT TAB(8);"(1/2)";TAB(18);"(1)";TAB(28);"(2)";TAB(38);"(5)"
    ;TAB(48)";TAB(58);"(20)"
 80 FOR W=0 TO K
 90 FOR P=0 TO K-W
100 FOR F=0 TO K-W(P+W)
110 FOR T=0 TO K-(P+F+W)
120 FOR O =0 TO K-(P+F+W+T)
130 FOR H=0 TO K-(P+F+W+T+O)
140 IF H+O+T+F+P+W<>K THEN 190
150 IF H*(0.5)+0+2*T+5*F+10*P+20*W<>J THEN 190
160 PRINT H,O,T,F,P,W
170 LET Z=Z+1
180 NEXT H
190 NEXT O
200 NEXT T
210 NEXT F
220 NEXT P
230 NEXT W
240 PRINT "All finished"
250 PRINT "The total number of solutions is";Z
260 END
```

Here is a typical print-out of the program:

```
Number of coins?
?8
Total amount of money?
?25

           (1/2)    (1)    (2)    (5)    (10)    (20)
           ------------------------------------------
             0       0      5      3      0       0
             0       3      1      4      0       0
             0       2      4      1      1       0
             0       5      0      2      1       0

All finished
The total number of solutions is 4
```

44 Cuisenaire Activity 6
Further rod equations

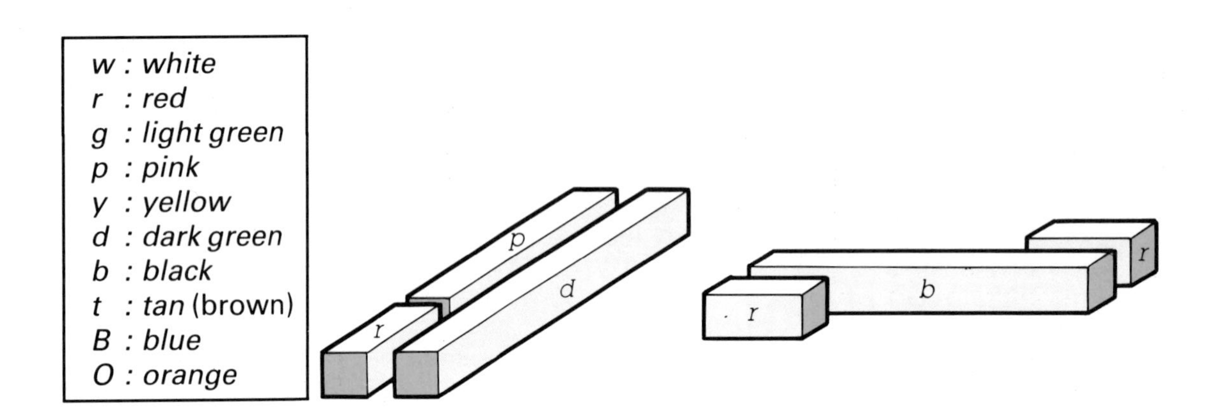

Find the 2 rods such that

$$F + S = g$$

and

$$F - S = w$$

where F = first rod, S = second rod.

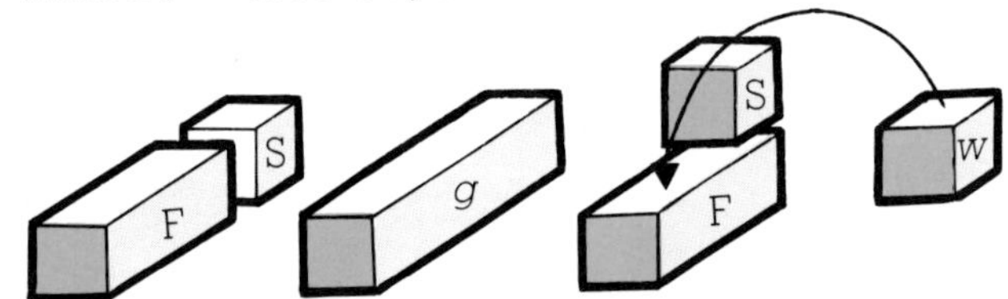

$F + S = g$ and $F - S = w$

Check that in both cases the solutions are $F = r$ and $S = w$.

Question 1

Find the two rods in each of the following such that:

1 $F + S = O$
$F - S = p$

2 $F + S = b$
$F - S = g$

3 $F + S = t$
$F - S = r$

4 $F + S = d$
$F - S = p$

5 $F + S = b$
$F - S = g$

6 $F + S = O$
$F - S = d$

7 $F + S = B$
$F - S = w$

8 $F + S = d$
$F - S = r$

9 $F + 2S = t$
$F - S = r$

10 $F + 2S = O$
$F - S = w$

Question 2

(Much harder)

1 $2F - 3S = b$
$F + 4S = B$

2 $3F + 2S = O + b$
$3F - S = y$

3 $F + S = b$
$F - S = r$

4 $2F + S = t$
$3F - 2S = y$

Explain as much as possible about how you have obtained your solutions.

Question 3

Using only *red* and *yellow* you cannot make a rod train equal in length to *dark green*. However, you can make any rod train longer than *dark green* so *dark green* is the longest length you *cannot make* using *red* and *yellow*.

What is the largest length you cannot make laying end to end, any combination of:

1 *green* and *yellow* rods?
2 *pink* and *yellow* rods?
3 *pink* and *tan* rods?
4 *pink* and *black* rods?

Generalise this result.

45 Number Theory 4
Further modulo arithmetic

Fractions in modulo

To interpret $\frac{1}{3}$ in Mod 7 we need to remember that

$$3 \times \frac{1}{3} = 1$$

So if we say

let $\quad x = \frac{1}{3}$ (Mod 7)

then $3x = 3 \times \frac{1}{3}$ (Mod 7)

$\quad 3x = 1$ (Mod 7)

Now in Mod 7, x can only be either of

0, 1, 2, 3, 4, 5 or 6.

So we set up a table

x	0	1	2	3	4	5	6
$3x$	0	3	6	2	5	1	4

When $3x = 1$ (Mod 7)
then $\quad x = 5$ (Mod 7)

but $\quad x$ was $\frac{1}{3}$ (Mod 7)

so $\frac{1}{3} = 5$ (Mod 7)

Question 1

Solve each of the following modulo equations. (In some cases there will be one answer, in other cases there will be more than one and in some cases there will be none.)

1 $3x = 1$ (Mod 5)
2 $3x = 2$ (Mod 7)
3 $3x = 1$ (Mod 6)
4 $5x = 3$ (Mod 7)
5 $4x = 4$ (Mod 6)
6 $2x = 3$ (Mod 9)
7 $5x = 2$ (Mod 7)
8 $4x = 1$ (Mod 5)
9 $3x = 0$ (Mod 6)
10 $2x = 5$ (Mod 9)

Square roots in modulo

To work out $\sqrt{4}$ in Mod 7 we need to bear in mind that

$$\sqrt{4} \times \sqrt{4} = 4$$

So if we set $x = \sqrt{4}$ then $x \times x = 4$ or $x^2 = 4$. In Mod 7, x can only be either of

0, 1, 2, 3, 4, 5 or 6.

So we set up a table again

x	0	1	2	3	4	5	6
x^2	0	1	4	2	2	4	1

When $x^2 = 4$ (Mod 7)
then $x = 2$ or 5 (Mod 7)
So there are two values of $\sqrt{4}$: these are 2 or 5 (Mod 7).
So $\sqrt{4} = 2$ or 5 (Mod 7).

We could also do this problem by saying:

$$\sqrt{4} = 2 \text{ or } -2 \text{ (Mod 7)}$$

because $2 \times 2 = 4$ and $-2 \times -2 = 4$.
So $\sqrt{4} = 2$ or -2 (Mod 7)
but $-2 = 5$ (Mod 7)
So $\sqrt{4} = 2$ or 5 (Mod 7)

As a third method we could proceed as follows:

$4 = 11 = 18 = 25 = 32 = 39 = 46 = 53 = 60 = 67 = 74 = 81 = 88 \ldots\ldots$ (Mod 7)

Taking the square roots of these numbers we see that 4, 25, 81 and some others have whole-number square roots of 2, 5, 9 etc. However, 9 is greater than 7 and $9 = 2$ (Mod 7). So the sequence repeats itself from here onwards. Hence

$$\sqrt{4} = 2 \text{ or } 5 \text{ (Mod 7)}$$

Question 2

Try these but be careful. In some cases there are no answers and in others there are more than 2.

1 $\sqrt{1}$ in Mod 7
2 $\sqrt{2}$ in Mod 7
3 $\sqrt{4}$ in Mod 5
4 $\sqrt{3}$ in Mod 6
5 $\sqrt{4}$ in Mod 6
6 $\sqrt{1}$ in Mod 8
7 $\sqrt{0}$ in Mod 9
8 $\sqrt{5}$ in Mod 9
9 $\sqrt{7}$ in Mod 9
10 $\sqrt{2}$ in Mod 4

Using the table method for solving modulo equations such as

$$5x = 2 \text{ (Mod 7)}$$

can be long-winded and messy for some numbers. There is a shorter, but harder method. We will look at an example. To solve

$$5x = 2 \text{ (Mod 7)}$$

we look at our 5 and 7 times tables. We have:

5, 10, 15, 20, 25, 30, 35, 40, ...

7, 14, 21, 28, 35, 42, ...

and note that 15 is 1 more than a multiple of 7. So

$$15 = 1 \text{ (Mod 7)}$$

This is our clue.

$$5x = 2 \text{ (Mod 7)}$$

We multiply by 3

$$15x = 6 \text{ (Mod 7)}$$

and because 15 = 1 (Mod 7)

$$x = 6 \text{ (Mod 7)}$$

which is our answer.

Question 3

Now try these using the shorter method (some can be difficult):

1 $3x = 1$ (Mod 5)
2 $4x = 3$ (mod 7)
3 $4x = 3$ (Mod 9)
4 $2x = 5$ (Mod 9)
5 $5x = 3$ (Mod 7)
6 $3x = 2$ (Mod 8)
7 $7x = 1$ (Mod 15)
8 $11x = 3$ (Mod 17)

Do not be afraid to experiment with these. Play with the numbers. See how this one below develops.

$$6x = 1 \text{ (Mod 13)}$$

Multiply by 2

$$12x = 2 \text{ (Mod 13)}$$

and since $12 = -1$ (Mod 13), we can say

$$-x = 2 \text{ (Mod 13)}$$

and multiplying by -1,

$$x = -2 \text{ (Mod 13)}$$

So, since

$$-2 = 11 \text{ (Mod 13)}$$
$$x = 11 \text{ (Mod 13)}$$

Question 4

A linear, modulo equation of the type

$$ax = b \text{ (Mod } n)$$

can have 0, 1 or many solutions. Solve each of the equations, by any means you wish.

1 $3x = 4$ Mod 8
2 $5x = 1$ Mod 7
3 $5x = 2$ Mod 6
4 $3x = 2$ Mod 6
5 $4x = 2$ Mod 6
6 $6x = 1$ Mod 8
7 $6x = 4$ Mod 8
8 $4x = 7$ Mod 8
9 $3x = 2$ Mod 9
10 $3x = 6$ Mod 9

From your results to these equations, and any others you have solved, obtain the conditions connecting the numbers a, b and n for equations of the type

$$ax = b \text{ (Mod } n)$$

to have:

(i) only one solution
(ii) more than one solution
(iii) no solution.

Offer any explanations you can as to why these conditions should be so.

46 Quadratics
Zeros

Suppose in the game of Frogs (see chapter 3) we have equal numbers of black and white counters. Suppose also that the game is completed in 35 moves. How many counters of each colour do we have?

In order to find the answer, we need to solve what is called a **quadratic equation**.

If we have x black counters and x white counters, the total number of moves is

$$\left.\begin{array}{l} x^2 + 2x \\ \text{or } x(x+2) \\ \text{or } (x+1)^2 - 1 \end{array}\right\} \text{these are some of the } \textit{generalised} \text{ results}$$

We could take any one, but will just choose $x^2 + 2x$.

If you are told that it takes 35 moves, then you must have:

$$\left.\begin{array}{l} x^2 + 2x = 35 \\ \text{or } x^2 + 2x - 35 = 0 \end{array}\right\} \text{each of these is called a quadratic equation.}$$

The expression

$$x^2 + 2x - 35$$

is called a **quadratic form**.

It has 1 lot of x^2
it has 2 lots of x
and it has a **constant term** of -35.

It is common to say:

the coefficient of x^2 is 1
the coefficient of x is 2
the constant term is -35

We can put values in for x. For instance if we put the value $x = 6$ into the equation, we get:

$$\begin{aligned} & 6^2 + 2(6) - 35 \\ =\ & 36 + 12 - 35 \\ =\ & 13 \end{aligned}$$

If we put $x = 3$ into it we get

$$\begin{aligned} & 3^2 + 2(3) - 35 \\ &= 9 + 6 - 35 \\ &= 15 - 35 \\ &= -20 \end{aligned}$$

If we put $x = 5$ into it we get

$$\begin{aligned} & 5^2 + 2(5) - 35 \\ &= 25 + 10 - 35 \\ &= 0 \end{aligned}$$

Because $x = 5$ makes $x^2 + 2x - 35$ worth 0, we say that:
5 is a **zero** of the quadratic form

$$x^2 + 2x - 35$$

Question 1

Check that $x = -7$ is also a zero of the quadratic form

$$x^2 + 2x - 35$$

Each quadratic form has 2 zeros. For the moment we are only going to look at quadratic forms whose coefficient of x^2 is 1. For example:

$$x^2 + 2x - 35$$
$$x^2 + 7x + 12$$
$$x^2 - 8x + 15$$

Your task is to work out the relationship which exists between the zeros of a quadratic form, its coefficient of x and its constant term.

Question 2

Check that the zeros of the following quadratic forms are correct.

	quadratic form	*zeros*
1	$x^2 - 8x + 15$	3 and 5
2	$x^2 - 9x + 14$	2 and 7
3	$x^2 + 2x - 15$	-5 and 3
4	$x^2 - 4x - 21$	-3 and 7
5	$x^2 + 7x + 12$	-3 and -4
6	$x^2 - 6x + 8$	2 and 4

Show your working. You may already have noticed something. Discuss it with neighbours.

Question 3

You are given 6 quadratic forms and told that their zeros are within a certain range of values. Find the zeros.

	quadratic form	*range of zero values*
1	$x^2 - 10x + 21$	between 0 and 10
2	$x^2 - 4x + 3$	between 0 and 6
3	$x^2 + 7x + 10$	between -7 and 0
4	$x^2 + 2x - 3$	between -4 and 2
5	$x^2 - 2x - 3$	between -2 and 4
6	$x^2 - 8x + 12$	between 0 and 10

If you have access to a computer you might be able to use it for the next question. You may use a calculator where appropriate.

Question 4

Record all of the results you have obtained so far. You might find it helpful to make a table like this:

quadratic form	*coefficient of x*	*constant term*	*zeros*
$x^2 + 2x - 35$	2	-35	5, -7

Can you give a **generalised** result connecting the zero's coefficient of x and the constant term?

Question 5

You are given the zeros. Write down the quadratic form.

1 4 and 8
2 -5 and -4
3 7 and -2
4 -3 and -6
5 9 and -12
6 5 and -4

Question 6

Find, as quickly as you can, the zeros of each of the quadratic forms:

1 $x^2 + 6x + 8$
2 $x^2 + 11x + 30$
3 $x^2 - 11x + 28$
4 $x^2 - 3x - 40$
5 $x^2 + 3x - 40$
6 $x^2 + x - 20$

Question 7

Try to find the zeros of each of the quadratic forms in which the coefficient of x^2 is something other than 1.

1 $2x^2 - 13x + 15$
2 $3x^2 + 7x + 2$
3 $2x^2 + 11x + 15$
4 $3x^2 + 17x + 10$
5 $5x^2 - 19x + 12$
6 $10x^2 - 19x + 6$

Can you find a **general** relationship between the zeros, the coefficient of x^2, the coefficient of x and the constant term?

Write up your findings in the form of a report.

Question 8

Write down the quadratic forms which have the following zeros in such a way that none of the coefficients are fractions:

1 $-\frac{1}{2}$ and 3
2 $\frac{1}{2}$ and -3
3 $-\frac{1}{2}$ and $-\frac{1}{3}$
4 $\frac{1}{5}$ and 2

47 Overlaps

Question 1

By overlapping the following pairs of shapes, how many other shapes can you make. Some suggestions are offered to help you.

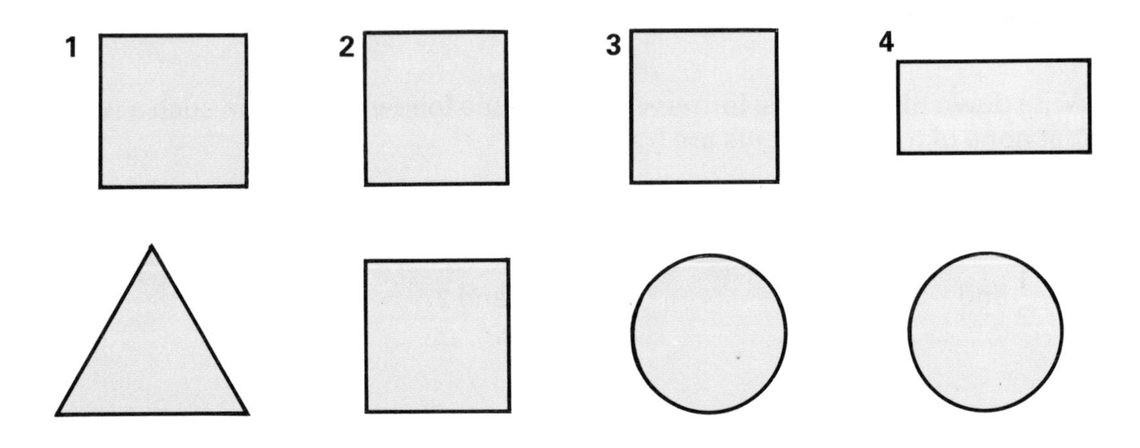

a square
a rectangle
a kite
a scalene quadrilateral
a triangle
(equilateral/isosceles/scalene)
a pentagon
a hexagon
an octagon

Are there other shapes you can make?

Question 2

Using thin sheets of paper make as many different overlap shapes as you can using any combination of the following four shapes. Record all the new shapes you make.

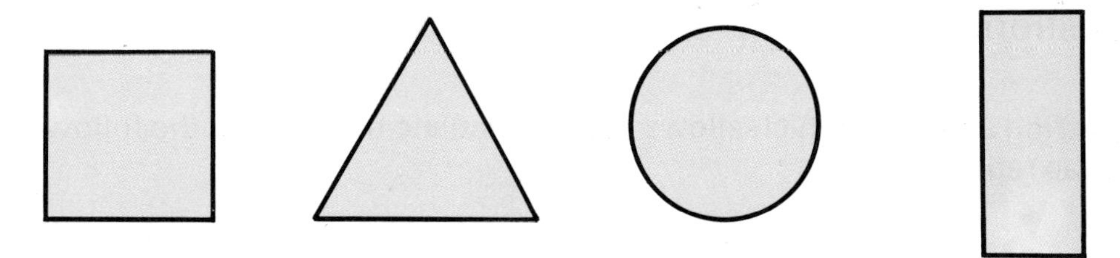

Example

The overlap of two different shapes produces a new shape.

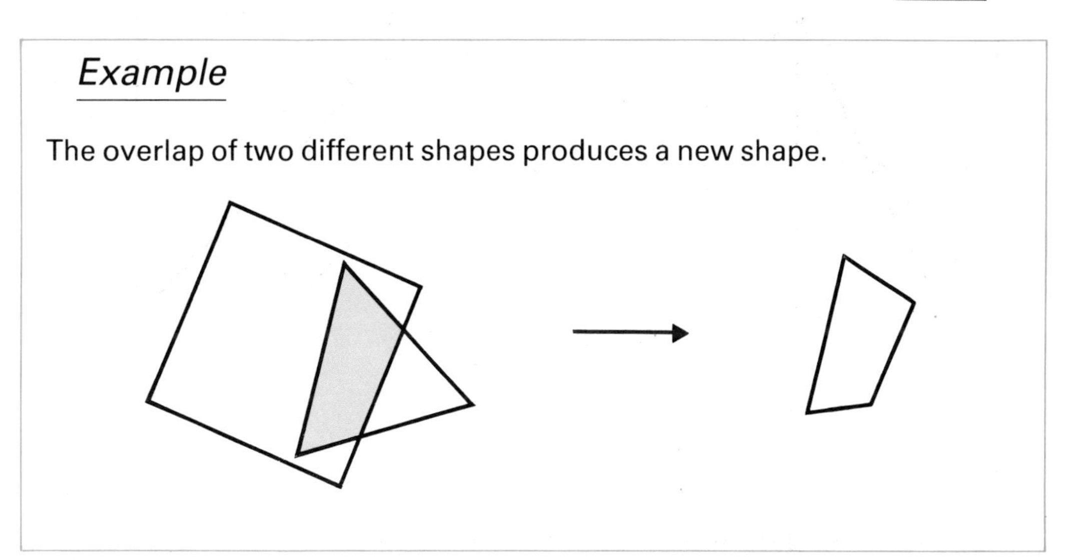

Question 3

Can you tell which shapes have been used to produce the following overlap shapes? Try to explain the observations which lead you to your conclusions.

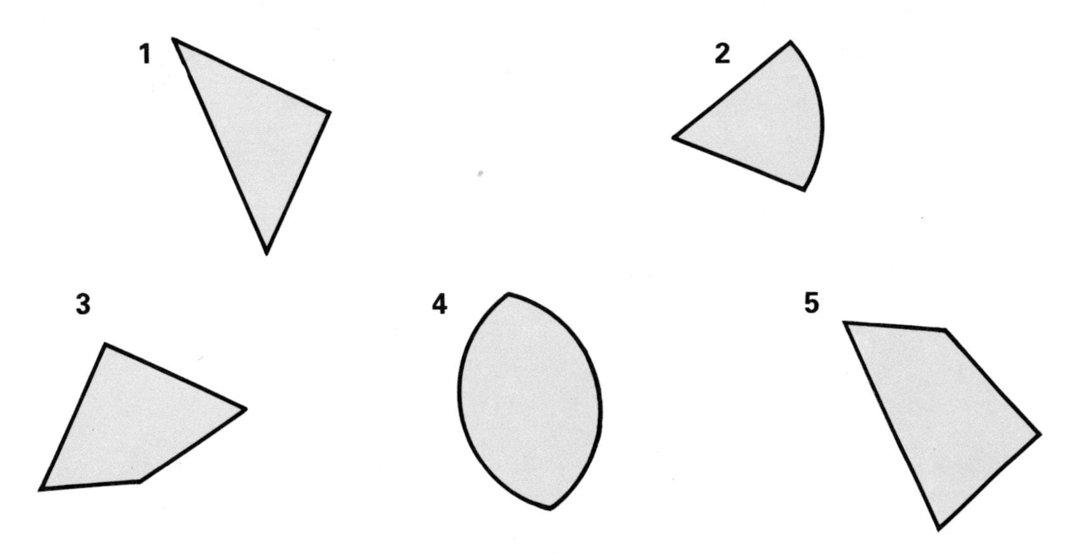

Question 4

Try to find a method which allows you to calculate the areas of the following overlap regions:

1

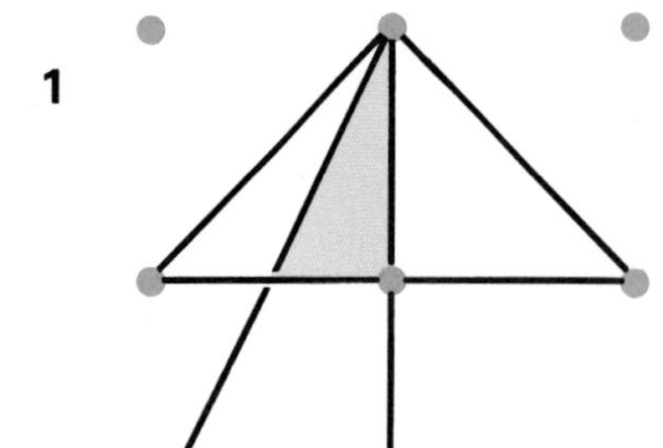

2

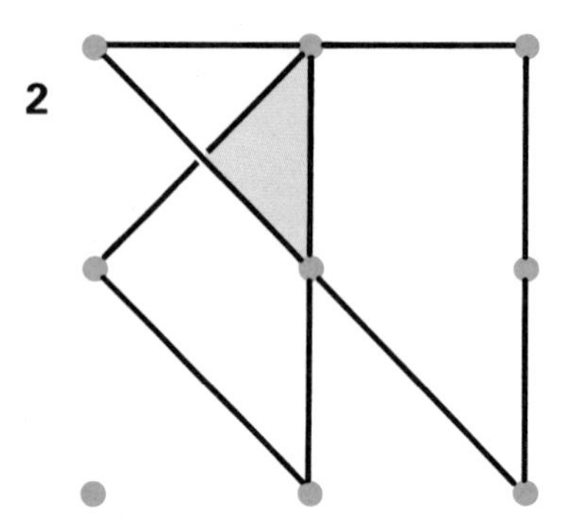

3

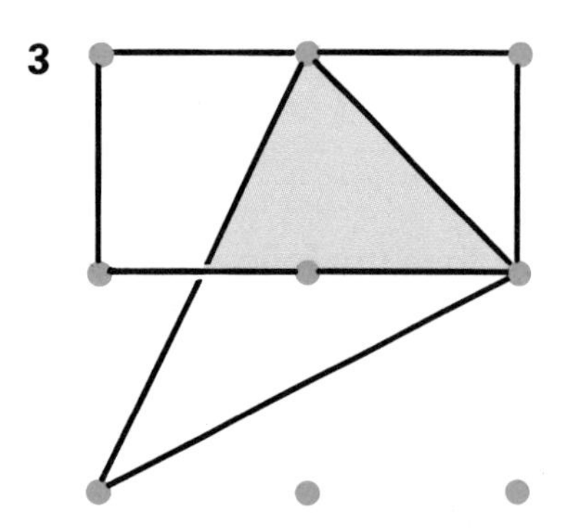

4

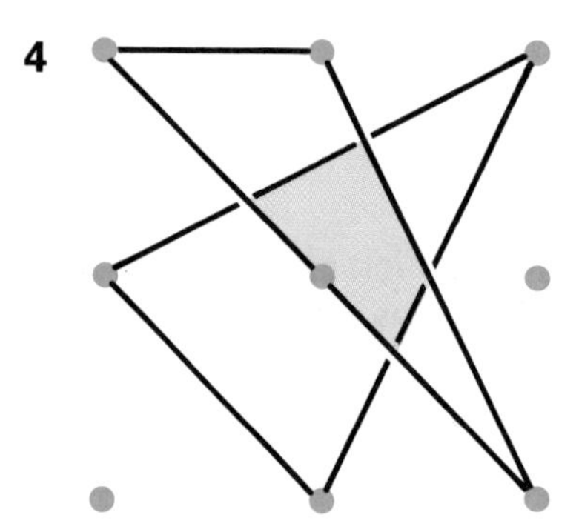

Now try some more of your own.

Question 5

Find some solid shapes produced using 3-dimensional overlaps.

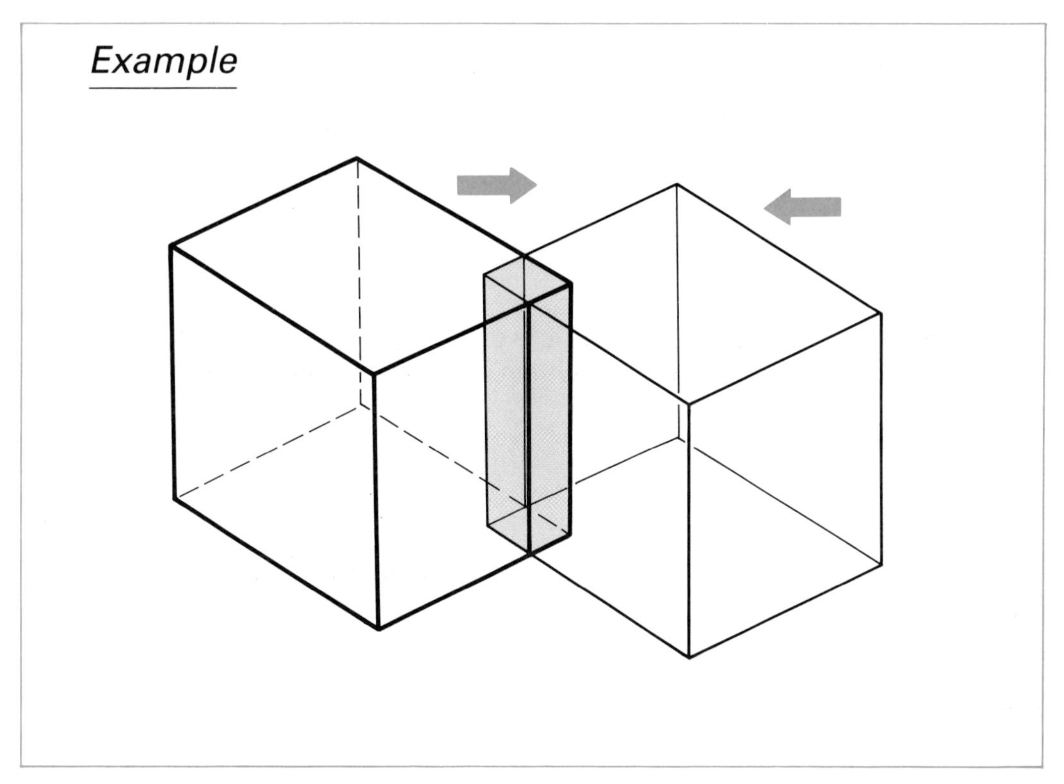

How could the cubes produce these overlaps?

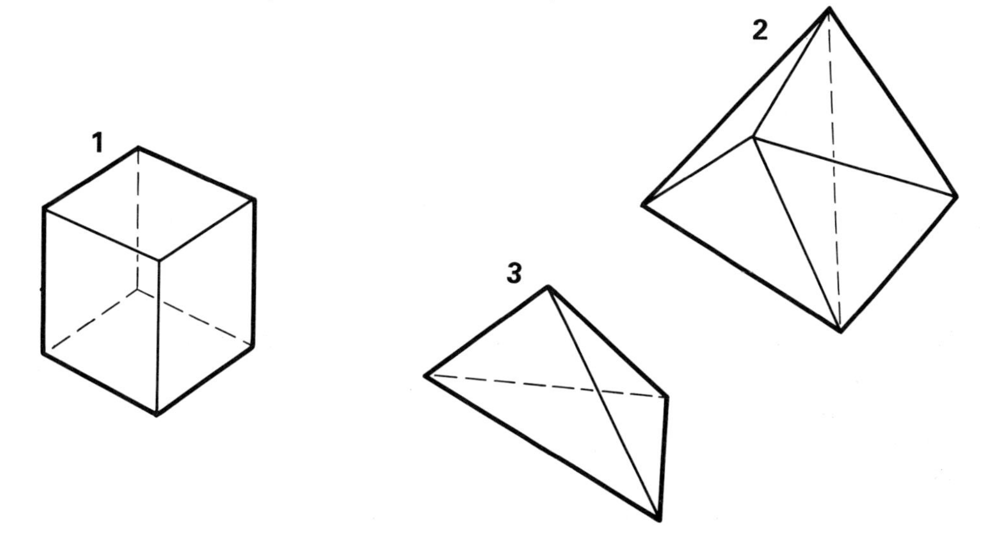

48 Bubbles

Here is an arrangement of 3 bubbles.

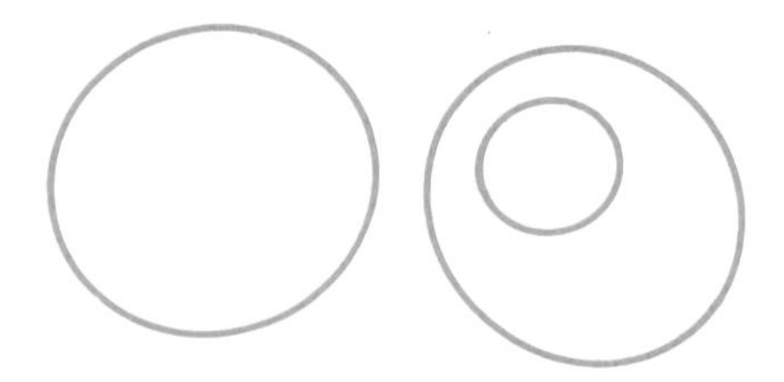

The bubbles can be arranged in a variety of ways. They can be separate, inside one another, inside a bubble which is inside another bubble and so on. However, they must not overlap.

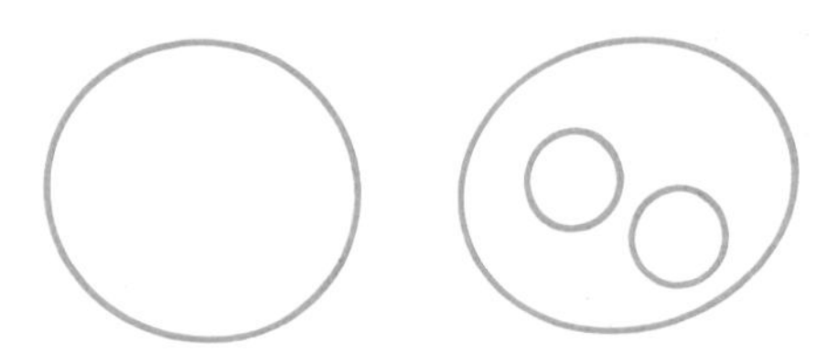

Here is an arrangement of 4 bubbles.

Size does not matter nor the order of the bubbles so

is the same as

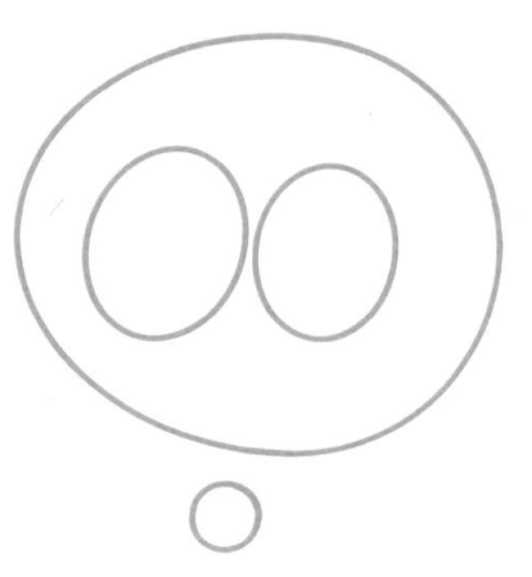

Question 1

(i) Three bubbles
How many different ways can 3 bubbles be arranged? Draw the possibilities.

(ii) Four bubbles
They can be arranged in 9 ways. Can you draw them all?

(iii) Five bubbles
They can be arranged in 20 ways. Can you find the different ways?

(iv) It helps to use a code to describe the arrangements. Can you use numbers to describe the arrangements? Try to invent a code.

Question 2

Here are some codes that pupils have suggested:
Code 1 (with 4 bubbles).
Look at the amount of overlap of the bubbles.

Can you see any problems with this coding system?

Question 3

Code 2 (with 4 bubbles).
Each bubble is given 3 points.
If a bubble is outside . . . . . you add.
If a bubble is inside . . . . . you multiply.

$3+3+3+3 \rightarrow 12$

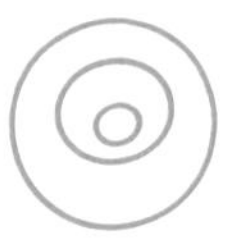

$3+(3\times3\times3) \rightarrow 30$

$3\times(3+(3\times3)) \rightarrow 36$

What numbers do you get for each of the other 4-bubble arrangements shown in question 1?

This code gives different numbers for each arrangement. Is it a good code?

Can you find a 5-bubble arrangement with codes of:

1 33
2 21
3 24
4 30
5 27
6 39

Is it easy to work back to the arrangement from the code?

Question 4

Code 3 (with 4 bubbles) Bubble brackets

1 — for a bubble on its own
1(1) — for a bubble inside a bubble.
1(1(1)) — for a bubble inside a bubble, inside a bubble.

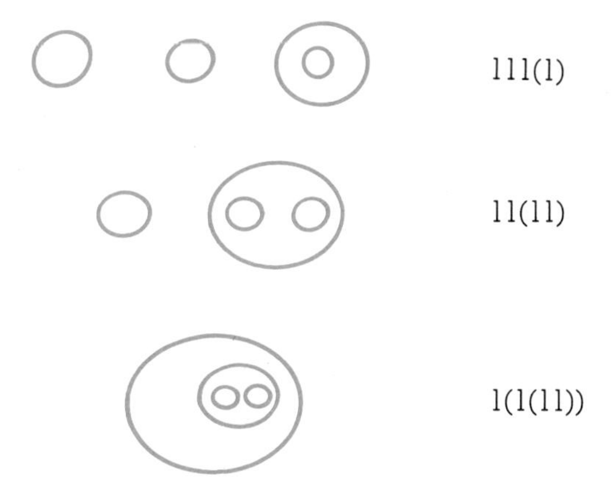

Give the codes for the other 4-bubble arrangements.

Here are the codes for 5-bubble arrangements. Can you draw arrangements given the codes.

11111	1111(1)	111(11)	11(1)(11)	1(1111)
		111(1(1))	11(1)(1(1))	1(111(1))
		111(1)(1)	11(111)	1(11(1)(1))
			11(11(1))	1(11(11))
			11(1(11))	1(11(1(1)))
			11(1(1(1)))	1(1(111))
				1(1(1(11)))
				1(1(11(1)))
				1(1(1(1(1))))

Question 5

In this question, find all the arrangements of 6 bubbles. Remember to be systematic in your approach.

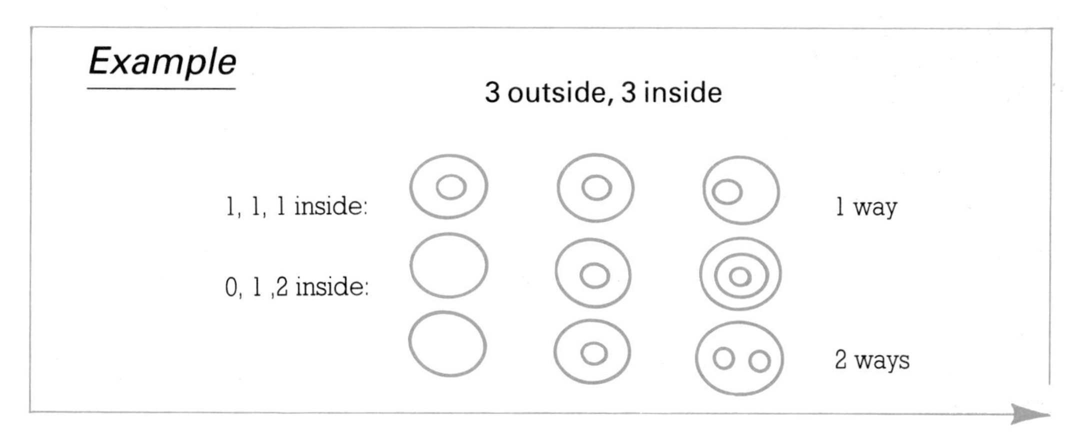

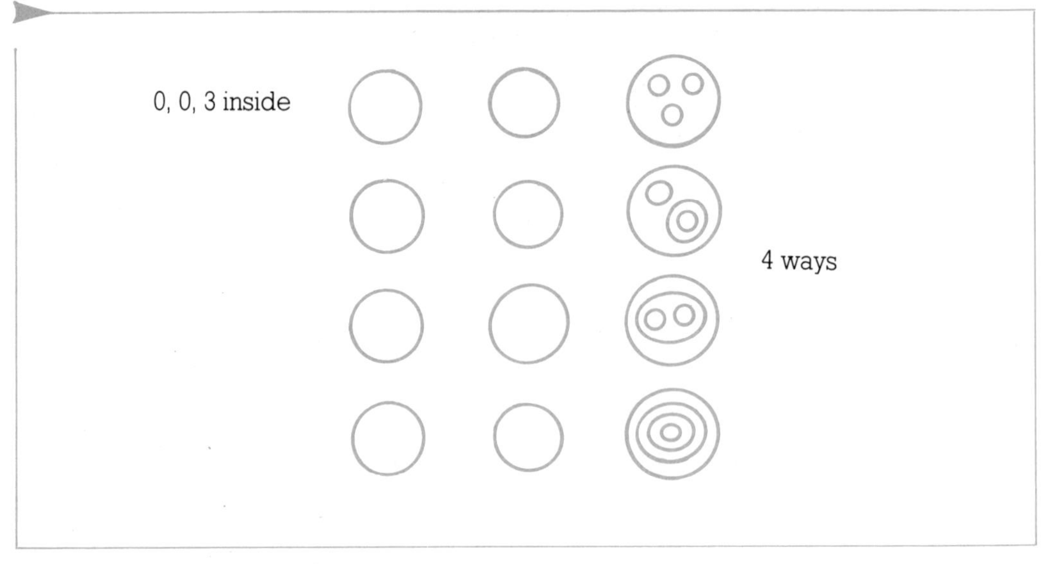

In the same way, look at:

1 6 outside, 0 inside
2 5 outside, 1 inside
3 4 outside, 2 inside
4 3 outside, 3 inside
5 2 outside, 4 inside
6 1 outside, 5 inside

How many arrangements are there altogether for 6 bubbles?

Question 6

7 bubbles
Find all the possible arrangements for 7 bubbles. Look at:

1 7 outside, 0 inside
2 6 outside, 1 inside
3 5 outside, 2 inside
4 4 outside, 3 inside
5 3 outside, 4 inside
6 2 outside, 5 inside
7 1 outside, 6 inside

Question 7

8 bubbles
Use these methods to work out the total number of arrangements for 8 bubbles.